Science Networks. Historical Studies

Birkhäuser

Science Networks. Historical Studies
Founded by Erwin Hiebert and Hans Wußing
Volume 60

More information about this series at http://www.springer.com/series/4883

Luca Guzzardi

Ruggiero Boscovich's Theory of Natural Philosophy

Points, Distances, Determinations

Birkhäuser

Luca Guzzardi
Dipartimento di Filosofia
Università degli Studi di Milano
Milano, Italy

ISSN 1421-6329 ISSN 2296-6080 (electronic)
Science Networks. Historical Studies
ISBN 978-3-030-52092-2 ISBN 978-3-030-52093-9 (eBook)
https://doi.org/10.1007/978-3-030-52093-9

This book is published under the imprint Birkhäuser, www.birkhauser-science.com, by the registered company Springer Nature Switzerland AG
The registered company address is: Gewerbestrasse 11, 6330 Cham, Switzerland

To my father
and to the memory of my mother

. . . We could not see without inventing what we saw, so at least we could try to do it properly. And then, because she shrugged dismissively and said: Why? Why should we try, why not just take the world as it is? I told her . . . that we had to try because the alternative wasn't blankness—it only meant that if we didn't try ourselves, we would never be free of other people's inventions—Amitav Ghosh, The Shadow Lines

Acknowledgments

This book is the result of 10 years of research, during which I was generously supported by several institutions. I wish to thank the Astronomical Observatory of Brera in Milan, where I started working in 2005 as an editor and collaborator at the Commission for the National Edition of Ruggiero Boscovich's Works and Correspondence. The former directors of the Observatory, Tommaso Maccacaro and Giovanni Pareschi, together with Elio Antonello, the president of the Italian Society of Archeoastronomy, provided optimal conditions for me to work and strongly endorsed the project of the Boscovich Edition. Edoardo Proverbio, without whose efforts such a project could never have been initiated, was, for me, a mentor in the Boscovich studies and a guide in organizing the editorial work. I am in debt to the Department of Physics of the University of Pavia, which funded the study of Boscovich's early works in natural philosophy in 2012–2014 and made possible their publication as vol. 6 of the Boscovich Edition, as well as to the Department of Philosophy of the University of Milan, which, since 2015, has supported my intensive exchange activity and my frequent stays at other research institutions. The Berlin-Brandenburg Academy of Sciences and Humanities hosted me in the fall of 2018 for a research project on Leibniz and Boscovich, funded by the Deutscher Akademischer Austauschdienst (DAAD): I owe a debt of gratitude to both for having made a significant portion of this book possible. Incidentally, the DAAD also granted me a fellowship in the spring of 2008 at the Research Institute of the Deutsches Museum in Munich to investigate Boscovich's sources and his place in early-modern European science. In both places, I met invaluable researchers to whom I address my sincere thanks, in particular, to Ivo Schneider in Munich, a mentor in the history of science who taught me how much the development of mathematics has shaped and still shapes our culture and why it is worthwhile to be aware of this fact—and to Eberhard Knobloch and Harald Siebert in Berlin, who carefully read and commented upon chapters of this book; their generous advice on the history of astronomy, as well as on the Leibnizian science, rescued me from numerous blunders.

Over the years, I discussed many parts of this work—mostly when I did not know that I was, in fact, writing a book—with colleagues and friends. Their contributions in shaping my own views through suggestions and criticisms are so profound and deep-wired in these pages that it is difficult for me to single them out. I only hope that I have not forgetten anybody: Ugo Baldini, Claudio Bartocci, Fabio Bevilacqua, Fabrizio Bonoli, Andrea Del Centina, Paolo Casini, Vincenzo de Risi, Stefano Di Bella, Steffen Ducheyne, Vincenzo Fano, Rivka Feldhay, Alessandra Fiocca, Antonella Foligno, Lucio Fregonese, Giulio Giorello, Pietro Gori, Pierluigi Graziani, Niccolò Guicciardini, Roberto Lalli, Henrique Leitão, Ivan Malara, Alessandro Manara, Ivica Martinović, Gianfranco Mormino, Elio Nenci, Matthias Schemmel, Corrado Sinigaglia, Josip Talanga, Tzu Chien Tho, Hans Ullmaier, Matteo Valleriani, and Ido Yavetz. I am indebted to all of them, in different manners and for different reasons. I am also grateful to two anonymous reviewers for their suggestions and feedback. And many thanks to Amitav Ghosh for allowing me to use a sentence from his inspiring novel *The Shadow Lines* in the epigraph.

* * *

From Boscovich's correspondence, it emerges that he had a difficult, fiery temper—unfortunately, this is the only aspect of his personality that I share. So, the greatest "thank you" is for my beloved Anna, who has not only been my first and most important aid in this peregrination, but also tolerates my worst Boscovichean intemperance.

* * *

I was in the midst of finishing this book when my mother got sick and died. Up to that moment, I never realized the extent to which and at what cost she and my father had influenced my own work—not in regard to having taught me what I should know, but by leaving me free to gain any knowledge I wished. This book is dedicated to them, as a sign of gratitude for having fostered my will to knowledge through their respect for liberty.

Introduction

The Man for Wisdom's Various Arts Renown'd

The intellectual portraits of Ruggiero Giuseppe Boscovich are at least as numerous as the media that portrays his features. Among the latter, the most famous is perhaps the painting by Robert E. Pine, which traces back to Boscovich's 1760 trip in England. However, if one types "Boscovich" into the search box of an Internet browser and selects "images," numerous pictures of him in many different poses and situations will surface. His sometimes round, at other times thin face is directed toward the viewer or appears in profile on modern stamps, banknotes, medals, and publications, as well as on bags and t-shirts.[1] In addition, many different spellings of Boscovich's name have been employed thus far. The most usual form in the English-speaking world (as well as in French) is Roger Joseph Boscovich, whereas in Croatia and the other Slavonic territories, his name is frequently written as Ruđer Josip Bošković, according to orthographic standards that trace back to the first half of the nineteenth century. Moreover, he was christened with the Latin version of his name, Rogerius Iosephus, and nicknamed Ruge within his family, as his sister Anica's letters confirm (for more details, see Boscovich 2012c, 1–2). I adopt here the spelling Ruggiero Giuseppe Boscovich, which reflects how he signed his own letters beyond the familiar circle.

This abundance of representations and identifications vividly epitomizes Boscovich's polytrophic personality. Born in 1711 in the independent Republic of

[1] In 2011, on two different occasions during the celebrations of Boscovich's tercentenary in Italy and Croatia, I was presented with bags—one from the University of Pavia and the other from the city of Dubrovnik—bearing Boscovich's portrait. Sometime later, during a conference, I saw a colleague wearing a t-shirt with Boscovich's face. I envied him a lot, but I must admit that I have a weakness for the mug that features a drawing of Boscovich's curve, which I received as one of the participants in the Boscovich Conference held at Pavia in September 2011. To the interested reader, another exemplar of a Boscovich mug can be seen (and used) at the Max Planck Institute for the History of Science in Berlin.

Ragusa on the Dalmatian coast (now Dubrovnik, Croatia), he began attending the Ragusinum, the Jesuit college in his city, during childhood. As one of its most talented pupils, at the age of 14, he was encouraged to go to Rome and pursue a career there as a Jesuit of the Roman College. In the citadel of Catholic orthodoxy, he received the typical training of the Society of Jesus, accomplishing each stage of the *Ratio Studiorum*. As a gifted young mathematician, he was appointed the chair of mathematics of the college in 1740–1741—thus becoming a successor of such regarded scholars as Clavius, Grienberger, Maelcote, and Gottignies. Contemporaries highly regarded and welcomed his mathematical and astronomical works, but his publication record extends well beyond mathematics, astronomy, and natural philosophy. It encompasses fields as diverse as geodesy, both practical and theoretical optics, meteorology, hydraulics, building engineering, gnomonics, and even the science of antiquity and poetry (most of all, didactic or occasional). In addition, he is also the author of an interesting travel journal that recounts his return trip from Constantinople to Rome through Eastern Europe.

His scientific vocation and the turbulent events of the mid-1700s factored into his frequent journeys throughout Europe. He routinely accepted—willingly and gratefully at certain times, out of necessity and due to force majeure at others—a variety of positions in different countries for limited periods. He preserved his post as the *professor matheseos* of the Roman College until 1763–1764, but during his 30 years of activity in Rome, he often traveled throughout Italy or abroad. In the early 1750s, he wandered throughout the Papal State, along with his confrere Christopher Maire, as a means to prepare for a new detailed geographical map commissioned by Pope Benedict XIV. Other travels, associated with various scientific interests or diplomatic tasks, included destinations such as Vienna (1757–1758), England and continental Europe (1760–1761), and Constantinople (1761–1762). The courts he visited in those circumstances often requested his expertise and advice on technical matters. Upon his return to Rome, the Senate of the Habsburg Milan appointed him professor of mathematics at the University of Pavia; he welcomed the new position and began his course of lecture in 1764, subsequently contributing, in the early 1770s, to the foundation of the Astronomical Observatory of Milan in Palazzo Brera, the seat of the Jesuits in the city.

Mala tempora currunt, however. In August–September 1772, as a consequence of the continuous fight between Boscovich and his confreres at the Brera Observatory, he was removed from his responsibilities as an astronomer, and resigned from the Pavia professorship some months later. He initially thought of going to Poland and visiting his native Ragusa. But ultimately—mostly due to the suppression of the Jesuit Order (1773), which rapidly extended from the European countries to the Papal State—some of his friends convinced him to go to France, where a post as the Director of Naval Optics of the French Navy was created for him. He returned to Italy in the mid-1780s to oversee the publication of his *Opera pertinentia ad opticam et astronomiam* (Bassano del Grappa: Remondini 1785); by this time, an agreement with the Habsburg administration, one that included bringing him back to Milan, seemed possible. His tasks might possibly have included cooperation with the astronomers of the now well-established observatory of Brera, who were charged with drawing a new map of Lombardy. However, his delicate health worsened soon after arriving there. He died on February 13, 1787, and was buried at the Church of Santa Maria Podone in Milan.[2]

* * *

My effort in this book is not for the purpose of bringing unity back to a wandering existence, nor making the intellectual life of its subject, which cut across a number of interests and disciplines, seem less nomadic. On the contrary, I believe that multiple, differing images of Boscovich are possible and that they are equally legitimate. The portrait I want to draw is that of "Boscovich the natural philosopher." Of course, this can be superimposed (and, in this study, *will* be superimposed) over other images of his activities, but it is also endowed with peculiarities that I hope will emerge on their own throughout the book.

The attempted portrayal, however, is only seemingly unambiguous. In fact, it requires that one at least approximately know what "natural philosophy" is. As remarked by Blair (2006, 365), in order to avoid appearing anachronistic, historians of science often use this phrase "as an umbrella term to designate the study of nature before it could easily be identified with what we call 'science' today." But this exact use involves some elusive degree of anachronism. As Blair continues, "'natural philosophy' (and its equivalents in different languages) was also an actor's category, a term commonly used throughout the early modern period and typically defined quite broadly as the study of natural bodies." It cannot simply be employed as an intemporal historical category, because it *was* a category on its own. It is, therefore, highly context dependent. Reacting against a possible unhistorical (as well as unphilosophical) adoption of this label, I agree with Schaffer (1980, 72) that "natural philosophy" cannot be taken as "a system of connected concepts and axioms with no definite associated form of practice," but should become a historical object itself and be understood as a form of (scientific) practice.

Of course, there is a simple and straightforward answer to the question "what is Boscovich's natural philosophy?" It is plainly all that which is contained in his fundamental book (1758, 1763) in this field, *Theoria philosophiae naturalis redacta ad unicam legem virium in natura existentium* (*A Theory of Natural Philosophy Reduced to a Single Law of the Forces Existing in Nature*). As a first approximation, this is almost tautologically true. We only have to single out its core and show its derivative concepts.

In a nutshell, the theory prescribes that every process in the world, beginning with those regarding the farthest stars down to those of the smallest particles, is an effect of a unique force that is repulsive at minute distances, attractive at large distances, and alternatively attractive and repulsive at intermediate scales. More exactly, bodies tend to move away from one another (repulsion) at very short distances, with a

[2]Two biographies of Boscovich are available to the English reader: Hill (1961) and Marković (1973, as part of the *Dictionary of Scientific Biography*). The comprehensive study by Marković (1968), in two volumes, is unfortunately only available in Croatian. Other two valuable biographical studies have appeared in Italian: Casini (1971) and Paoli (1988).

repulsive force that grows to infinity when the gap dividing them is infinitely small. However, they tend to move toward one another (attraction) when the interval involved is large enough (e.g., at planetary distances), so that it approximates Newton's gravitational law. Between these two extremes (infinitesimal distances—planetary distances), bodies alternatively approach and move away from one another depending only on the distances at which they are posed. Boscovich argued that, at distances greater than those characterizing the solar system, the force could become repulsive again, so that the stability of the system is ensured.

The idea of the curve was first introduced in a 1745 dissertation, *De viribus vivis* (*On the living forces*), in order to avoid the abrupt changes of velocity that were prompted, according to Boscovich, by the Cartesian treatment of collision. For Descartes and his devotees, Boscovich contends, velocity changes—hence motions—are generated through collisions. However, he observed that the Cartesian rules of impact allow for an immediate inversion of velocity in the instant of collision:

> Let us assume—as in (Fig. 1)—that two identical elastic spheres *AB*, *CD*, moving with equal velocities (which are expressed by the lines *AF*, *DO*, perpendicular to *AD*) collide in *E*. In the same instant of time when the points *C*, *B* of the diameters touch one another, they necessarily arrest their motion, while the diameters *BA*, *CD* end up in *Ea*, *Ed*, equal to one another. However, all the remaining particles except of the first ones, but including the last ones (*a* and *d*), keep on moving with always reducing speeds, until their velocity are entirely extinguished in *M* and *N*, now with changed form and shortened diameters. If the spheres would be soft, they will preserve this state; if they would be elastic, the individual particles will bump back with the same degree of velocity. Let us keep on drawing the perpendiculars *BG*, *aH*, *EI*, *dK*, *CL* up to the line *FO*, then the velocities of the points *A* and *D* will obviously be expressed through ordinates to *FO*, which ordinates will be equal to one another until *H* and *K*; then through ordinates to the endlessly decreasing lines *HM*, *KN*. But the velocities of the particles *B* and *C* . . . would totally be extinguished in an instant of time and [. . . they] would be at rest for all the continuous time *a* and *d* would take to reach *M* and *N*. So, those velocities are given through ordinates referred to the line *FO* until *I*; then in *I* every

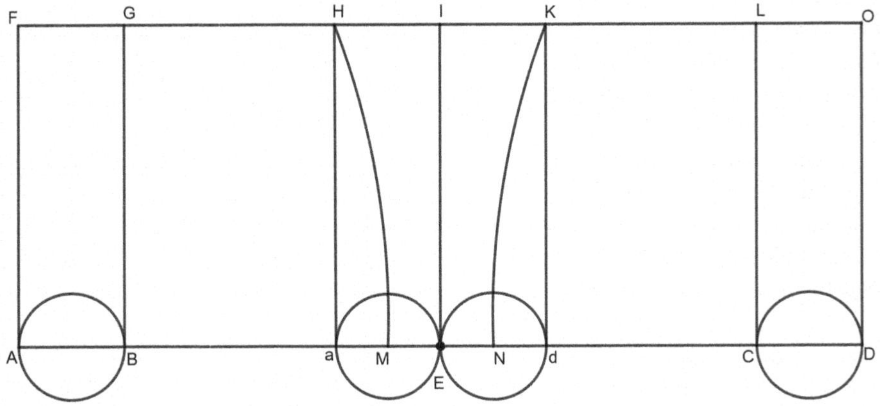

Fig. 1 Diagrammatical representation of the supposedly instantaneous annihilation of velocity according to Boscovich (Modern adaptation from Table I, Fig. 9, of Boscovich 1745)

> expression by means of the ordinates breaks down, and the ordinate *EI* will be followed by a point. (Boscovich 1745, § 46)

In brief, if Descartes was right, then the velocities would abruptly be destroyed and immediately re-created at the instant of contact. But this was something that Boscovich was not willing to accept. On the contrary, he proclaimed that instantaneous changes in the state of motion of bodies should be avoided as contrary to nature, for "it is already widespread the judgement of many people, that in nature nothing happens by jump" [*Communis iam est multorum sententia, nihil in natura per saltum fieri*] (Boscovich 1745, § 45). Continuity of natural processes, Boscovich insinuates, must simply be assumed as real (as I shall remark in Chap. 5, this continuist position was a result of his strong commitment to the Aristotelian doctrine of continuity.) Conversely, there is a version of the Newtonian theory of distant forces that is able to preserve the continuity of processes: there should be a certain mechanism according to which bodies slow down when approaching one another and pass through all velocity degrees before stopping, changing, or inverting their directions. Such a mechanism was explained in terms of a repulsive force (determination) acting at a short distance and growing when the distance lessens:

> Let us assume that no velocities can be extinguished in an instant of time. [In this case] the spheres do not keep on moving with uniform velocity until contact, but whenever particles *B* and *C* would arrive at a very little distance, some repulsive force will push them back endlessly, so that their velocities will gradually be extinguished before contact. By replacing rigid bodies with soft or elastic bodies, there will be no jumps in the velocities of particles *A* and *D*. But the jump in the velocities of particles *B* and *C* cannot be avoided, unless such repulsive force will be assumed at the smallest distances. (Boscovich 1745, § 47)

In the following 10 years or so, Boscovich would develop this early concept of a force law and enhance it with other elements, such as the investigation of matter composition, the law of continuity that he had used since 1745 as a heuristic principle, and the development of adequate mathematical tools. The final formulation of the law of forces would appear, in the abovementioned book of 1758/1763, in the form of an elegant diagram that vividly represents the most important features of the theory. These can be described in qualitative terms as follows: The curve expresses a force acting between two points, one at the origins of the axes in A and the other moving along the x-axis. With the distances plotted on the x-axis, the force intensity is represented on the y-axis. It is a continuous quantity that varies with distances: at each distance, there is only one possible force value (or magnitude or degree); if distance is only slightly varied, then force is also varied. In this sense, the force is a continuous function of the distance. Repulsion and attraction are not different forces, but rather different phenomenal manifestations of the same force. Looking at the graph (Fig. 2), force is repulsive when the curve is above the x-axis and attractive when it is under it. Moreover, it is infinitely repulsive when the distance is infinitely small (ED grows asymptotically), and infinitely small when the distance grows infinitely (more precisely, as explained in § 405 of *Theoria*, it is very small at planetary distances). In between, the force is alternatively repulsive and attractive, depending on distances, and, because force is conceived as a continuous

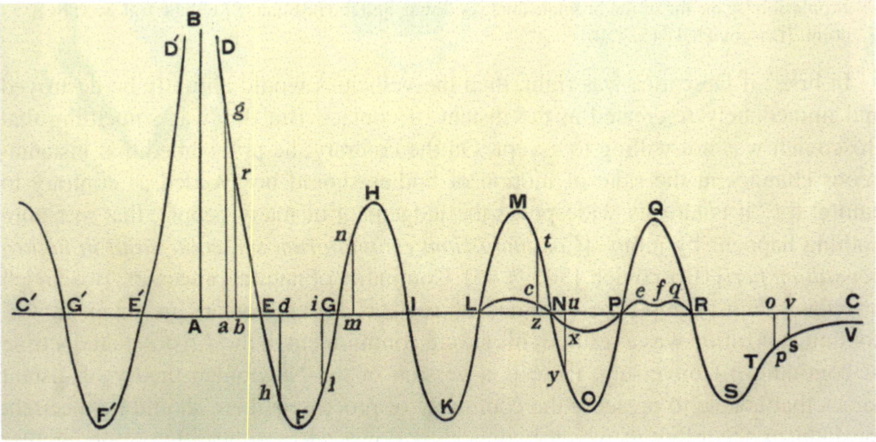

Fig. 2 Boscovich's curve. (From Ruggiero Giuseppe Boscovich, *A Theory of Natural Philosophy*, ed. by J.M. Child, New York: The Open Court, 1922). In truth, this is just a basic instantiation out of many possible shapes, as can easily be seen from the analytical expression of the force function that Boscovich published in one of the "Supplements" to the treatise (see *this book*, Sects. 4.3, 6.4, and 6.5)

function of x, there must be distances at which attraction turns into repulsion, and *vice versa*.

Determining the distances at which force inverts its "sign" or "direction" is a matter of experimental physics: as Boscovich already emphasized in one of his early works, finding the points at which inversions occur (i.e., the points at which the curve cuts the axis) is an empirical problem "to be investigated from the phenomena" (Boscovich 1745, § 56). Yet, he never proposed or performed any experiments, nor did he develop quantitative methods to prove his theory. Ultimately, an analytic solution complemented the graph of the curve: as we shall see in Chap. 6, this was a convoluted ratio between two polynomials, with an undetermined number of parameters.

By these means, Boscovich maintained, all natural phenomena are described: gravitation on the largest scale; the impenetrability of matter at infinitesimal distances (so that no actual contact takes place); cohesion as a balance between attraction and repulsion in the shortest range; "fermentation" (i.e., broadly speaking, chemical processes involving a transformation of one substance into another through the release of air from solids or liquids), and electricity and magnetism in the intermediate range, where both repulsive and attractive behaviors manifest themselves.

* * *

This is, roughly speaking, the kernel of Boscovich's book *Theoria philosophiae naturalis*. But, of course, his natural philosophy as a whole cannot be reduced to this—in the same sense that, for example, Newton's natural philosophy cannot be reduced to the Law of Universal Gravitation. In fact, most reconstructions of

Boscovich's natural philosophy so far have insisted that its main feature was a certain tendency toward unification. After all, the second part of the title of Boscovich's book on natural philosophy, with its emphasis on a "single law of forces," seems to announce that the author's intention was to provide a *unified* theory of forces. Accordingly, many discussions of Boscovich's theory have emphasized that, at the midpoint of the eighteenth century, researchers had available a crowd of forces associated with their carriers. Boscovich explicitly intended to unify that variety of interactions by gathering them into a single law that, in the end, associated every force with a definite interval between point-like particles. A plurality of forces; a unified framework.

Indeed, this image of Boscovich's natural philosophy was codified in the late nineteenth century as an effect of achievements made in physics. It was the theory of field developed by Faraday and Maxwell that unified diverse phenomena, which had up to then been viewed as distinct effects of different forces, into a single framework. The theoretical field conception, sanctified by Heinrich Hertz's 1888 experiments, primarily applied to electricity, magnetism, and optics—but their champions allowed for speculation that further extensions were possible. In light of this, they took Boscovich as their precursor toward unification. They interpreted his "points of matter" as point atoms or, more radically, as force centers; finally, Boscovich's force—like "an atmosphere ... grouped around" a mere mathematical point, as Faraday (1844, 290) put it—would be the ultimate reality, and no less than the ancestor of the almighty, all-embracing field. From this point on, Boscovich's natural philosophy has been interpreted, virtually with no exceptions, as a forerunner of the theory of field, force as a physical entity unto itself, and his points of matter as point atoms or force centers.

However, if one reads Boscovich's own presentation of his theory carefully, such an interpretation does not seem very accurate, to say the least. Let us examine an example of such a presentation:

> I ... admit that any two points of matter are subject to a determination to approach one another at some distances, and in an equal degree recede from one another at other distances. This determination I call force; in the first case attractive, in the second case repulsive; this term does not denote the mode of action, but the determination itself, whatever it comes from, of which the magnitude changes as the distances are changed. This determination follows a certain law, which can be exposed through a geometrical curve or an algebraic formula, and represented to the eyes, as it is usual to the scholars of Mechanics. (Boscovich 1763, § 9)[3]

[3]"*Censeo ... bina quaecunque materiae puncta determinari asque in aliis distantiis ad mutuum accessum, in aliis ad recessum mutuum, quam ipsam determinationem appello vim, in priore casu attractivam, in posteriore repulsivam, eo nomine non agendi modum, sed ipsam determinationem exprimens, undecunque proveniat, cujus vero magnitude mutatis distantiis mutetur et ipsa secundum certam legem quandam, quae per geometricam lineam curvam, vel algebraicam formulam exponi possit, et oculis ipsis, uti moris est apud Mechanicos repraesentari.*" Quotations from Boscovich's *Theoria* are mainly taken from J.M. Child's Latin-English edition (and are referred to as Boscovich 1763); some changes might have been introduced in the translations. In all other cases, translations are my own, unless they are otherwise noted. Quotations from

I will postpone a more detailed commentary of this and other passages, as well as a discussion of Faraday's contention about Boscovich and his theory, until subsequent chapters (see, in particular, Sect. 3.4.2). Here, I only want to remark that Boscovich is not saying that, in nature, there are many qualitatively different forces that are going to be unified by means of his single law or curve. If we take his statement at face value, as historians (and also philosophers, hopefully) are supposed to do, he is, rather, stating that every *couple* of *points* (and not every *individual* point, as Faraday's image of an atmosphere surrounding each point suggests) is subject to *one* mutual force—and always the same force—which would be better called a *determination*, that is, a measure of the propensity to perform a certain motion, irrespective of its cause (I will explain this notion in Chap. 2). This determination has two opposite expressions: two points can mutually be determined to approach one another (in which case, as Boscovich claimed, we call it a force of attraction) or they can mutually be determined to separate (in which case we call it a force of repulsion). The quality of the forces involved—chemical, magnetic, electric, gravitational, etc.—does not play any role, to the extent that it is not even mentioned, and to the extent that Boscovich added that the term force "does not denote the mode of action." Indeed, we are told that the only thing that matters is the *distance* between points: the determination of approaching or separating not only generically depends on distances; it is a function of the distance; its "magnitude changes as the distances change," according to a certain law that can be visualized by means of a curve, whose graph represents how force intensity varies over distances.

In view of this, it seems to me that a discussion of Boscovich's conception should not start with presupposing force unification as a reaction to a previous force proliferation, and should also avoid starting with the presumption that such a conception is a forerunner of whatever theory has later appeared (which does not mean, of course, that it had no effect on later developments in physics or other fields).[4] On the contrary, I think that a more adequate image of Boscovich's natural

---

Boscovich's works refer to the relevant paragraphs, introduced by the sign §. Where this sign is not present, the numbers are meant to refer to the relevant pages.

[4]So, the view expressed in this book is at odds with the now widespread tendency to read the *Theoria philosophiae naturalis* while taking on the background modern conceptions of matter constitution. This kind of modernizing reading is instantiated in the entry "Ruđer Bošković" of the *Hrvatska enciklopedija* ("Croatian Encyclopedia," vol. II, Zagreb: Leksikografski zavod Miroslav Krleža, 2000, 270–272), which has made available in English on the occasion of the Croatian celebrations for the 300th anniversary of Boscovich's birth. One of the most interesting (and, to me, astonishing) passages reads as follows: "A single law of forces existing in nature (lat. *lex unica virium in natura existentium*), i.e. the idea that one law can explain all of reality, constitutes Bošković's main contribution to science. The same idea has been entertained by A. Einstein, W. Heisenberg and more contemporary scientists, but the four forces in nature (gravitational, electromagnetic, weak and strong nuclear energy forces) have yet to be described by a unified theory. Bošković's single law is a framework for a unified theory of fields or, even more so, for a Theory of Everything." Other attempts within this approach have been more cautious. For example, cosmologist John Barrow (2007, first published in 1991) has presented Boscovich's theory as the first serious effort to attain unification of the fundamental forces governing nature, but not necessarily in the sense of a TOE. In addition, Ullmaier (2005) has offered a comprehensive and

philosophy should include an effort to explain how these three elements—*points* of matter, *distances* between points, and the *determinations* that arise from them—could possibly provide an efficient basis for it, at what cost, and with what consequences. This book, titled according to such a triptych, represents this kind of effort.

* * *

In another respect, this discussion of Boscovich's natural philosophy is different from the presentations made so far. Some of them are invaluable contributions to the history of science, but they offer an example of natural philosophy conceived as a *system of connected concepts and axioms*, to pick up Schaffer's phrase. They tend to consider natural philosophy as a stable form that can be filled in with changeable content: in the end, all of the natural philosophers do the same thing; they only disagree on how to do it. They can have different ideas regarding forces, matter, and their interactions, but the kind of connections among these concepts does not change.

For example, we learn from Mary Hesse's influential book *Forces and Fields* that Boscovich's theory of force as a continuous function of the distance between two points

> may be criticised on grounds of being *ad hoc*, but Boscovich was misled by his data rather than his method. Considering the confused state of the theory of matter and of chemical interaction at the time Boscovich wrote, it was not to be hoped that the details of his theory would survive, but his method was in the tradition of mathematical physics, leading from Newton to the present time, indeed the method of deriving a force-function *ad hoc* from the phenomena is very similar to that used at present in postulating short-range nuclear forces. (Hesse 1961, 165)

One could even imagine that, had he only known all of that which we now know about the structure of matter, he surely could have improved the content of his theory, because the method was sound. Indeed, he actually did what we are currently doing, that is, what we have always done in practicing physics—what else? It is, of course, of minor importance, for example, what Boscovich's institutional role was (if any), in which intellectual and material environment he was educated, with whom he shared ideas (and with whom he did not), or according to which criteria he presented his results, data, or conjectures. After all, theories are impersonal; once they are articulated, they live on as their own, as a crystal-clear texture of logical interrelations.

In this book, I will avoid this kind of Platonism. Beginning with Fleck's pioneering study *Entstehung und Entwicklung einer wissenschaftlichen Tatsache* (1935), it has been abandoned in practically all fields of the integrated history and philosophy of science, not to mention disciplines like social epistemology, the sociology of science, cultural and material studies, etc. So, I do not feel any need

---

valuable overview of the analogies and differences between Boscovich's insights and modern theories of matter.

to justify my dismissal, even if there are some pockets of resistance. As regards the Boscovich studies, I simply suggest turning things upside down: not to consider his natural philosophy as a system of *concepts*, but as a scientific *practice*—more precisely, as a result of a web of practices. So, the starting point (and, to some extent, the most important point) of this book will not be concepts, but the practice in which Boscovich was educated and to which he was supposed to contribute as the *professor matheseos* of the Roman College, the citadel of Jesuit orthodoxy: namely, mathematics. I will argue that Boscovich's theory was the outcome of a peculiar epistemic attitude that was deeply rooted in his mathematical education—and I will use the phrase "agnostic neutralism" or similar expressions to label this attitude.

Chapter 1 analyzes the mathematical tradition in which Boscovich grew up—the tradition of Cristophorus Clavius. It discusses the emergence and development, within the Roman College, of the so-called *physico-mathesis*, a noticeable propensity of the seventeenth- and eighteenth-century mathematicians to expand the scope and methods of their discipline and colonize physics. It emphasizes that the peculiarity of the Jesuit *physico-mathesis*, together with the obligation for the Order to defend a geostatic and geocentric cosmology (which remained in effect until 1757), drove the mathematicians of the Roman College to support geostatic-geocentric systems that were mathematically compatible with that of Copernicus. It argues that a persistent trait of Boscovich's epistemology, i.e., a certain *agnosticism* about the true intentions of God as the "Author of Nature" and a corresponding *neutralism* regarding the variety of physical interpretations that we can give to a mathematical law, is essentially based on this kind of compatibilism, to which he contributed in an original manner. (Sometimes, I will refer to Boscovich's approach as a *physico-mathematical style*, hinting at "his" mathematical tradition.)

Chapter 2 contends that Boscovich's agnostic-neutral perspective, along with its mathematical background, was instrumental in the development of his concept of force and guided him toward the formulation of the curve of forces. I argue that there is a strong relation between his agnosticism-neutralism and his considering forces as "mathematical determinations," as can be understood from a close examination of the early writings of mechanics, in which both inertia and external forces are defined in those terms. A comparison of the early texts with later works, such as the 1745 dissertation *De viribus vivis* and the 1758/1763 *Theoria*, in which those definitions are repeated, suggests that the notion of determination was the most important element in Boscovich's epistemology of force.

It is a reasonable expectation that different epistemic attitudes, possibly stemming from different practices, may also result in different conceptual constellations—even if superficial analogies can emerge from a sea of diversity and stand out as dominant features. Chapter 3 contrasts Boscovich's conception of force with other coeval and apparently similar approaches. He often emphasized that he was inspired by the originally Newtonian insight that, at certain distances, an inversion in the direction of force takes place, so that attraction turns into repulsion. In fact, this idea can easily be found in the final Query to the *Opticks* and, as often pointed out by historians, it was developed in the first half of the eighteenth century by "Newtonianizing" British physicists like Stephen Hales, John Theophilus Desaguliers, Gowin Knight, and John Michell. They are usually labeled as "dynamical corpuscularists" (e.g., by Schofield 1970), and it has often been claimed that their theories are an anticipation of Boscovich's. More precisely, the latter's natural philosophy would be none other than a variety of a "dynamical theory of matter" based on Newton's notion of active principle. I will discuss this view and argue that, once Boscovich's physico-mathematical style is taken into consideration, things are not as linear as they seem *prima facie* and I will advance a different framework for understanding the relationship between Boscovich and this group of British "Newtonians." This picture might also help reframe a couple of questions about his alleged influences. So, in the final section, I will discuss the two "vexed questions" of how his natural philosophy would be influenced by Leibniz's concept of force and how it might have influenced Faraday's idea of field.

To the present portrait of Boscovich as a natural philosopher, Chap. 4 adds strokes different in kind and coming from a different source. Based on his correspondence, I emphasize, and provide documentary evidence for, the influence of the Aristotelian tradition in the theory of matter involved in his conception of force. Several studies have investigated the role and the many nuances of Aristotelianism within the Society of Jesus (see, e.g., Crombie 1975, Dear 1987, Feldhay 1987, Baldini 1992b and 1998, Simmons 1999, Feingold 2003). To a mathematician trained in the Jesuit tradition, Aristotle surely had something important to offer: the analysis of continuum. I contend that Boscovich's notion of material point (or point of matter) as the elementary constituent of bodies—a typical *non*-Aristotelian doctrine—was the combined effect of his mathematical interests and the *Aristotelian* treatment of potential infinity as the outcome of an iterated operation of dividing a continuous quantity. This also provides the background for Boscovich's concept of mass and his conception of the material world as an aggregate of aggregates of different orders of size.

The problem of continuity in Boscovich's theory is specifically addressed in the subsequent Chap. 5. It has often been emphasized that he explicitly took continuity as a "metaphysical assumption" (see, e.g., Martinović 1987; Čuljak 1998, 2008; Talanga's Introduction to Boscovich 2001; Heilbron 2015)—a principle that he expressed with reference to Leibniz's law of continuity in order to justify the curve of forces. I will discuss how the role of continuity has changed and gradually acquired importance in the development of Boscovich's natural philosophy. In particular, starting with Aristotelian premises, he arrived at formulating a principle of continuity that introduced and reinterpreted elements from Leibniz's doctrines. However, in this process of adoption and adaption, a substantial role was played by a particular aspect of Boscovich's mathematical practice, namely, the use of geometric diagrams. (The title of the chapter takes inspiration from Leibniz's statement in the *Theodicy* about the "two famous labyrinths," the one concerning freedom and necessity and "the other consist[ing] in the discussion of continuity.")

Chapter 6 argues that the explicit articulation of the concept of continuity that found expression in *De continuitatis lege* (1754), together with the exploration of its mathematical underpinnings taking place in coeval works on conic sections, curves, and the philosophy of mathematics, guided Boscovich to the final refinement of the law of forces and to his search for its analytical formula. Here, I will pay particular attention to the mathematical constraints of the curve of forces and comment on the six fundamental conditions that, according to him, the formula is supposed to satisfy. Finally, I will give a global assessment of his natural philosophy in light of such strong mathematical commitment.

Contents

Chapter 1
In the Temples of Holy Mathematics

1.1 The Good Purposes of Father Clavius

In the Jesuit order of knowledge, mathematics played an important but controversial role, which was partly derived from and partly in tension with the Aristotelian-Scholastic tradition. This, in turn, provided the common framework for the diverse philosophical tendencies developed within the Society.[1] Jesuits tended to conceive of mathematics as a whole as being divided into two kinds: *mathesis pura* and *mathesis mixta*—a categorization that only partially matches up with the typical modern subdivision into "pure" and "applied" mathematics. As Bos (1980, 329) pointed out, in this context, "the object of mathematics is the mutual relations of magnitude and number of any objects which are capable of increase or decrease." The field of pure mathematics embraces "the relations between (variable or constant) quantities irrespective of the objects they measure or count," whereas mixed mathematics studies "quantities, and their relations, as they occur in natural objects which can be counted or measured."[2]

The partition was traditional, and traced back to Medieval and Renaissance mathematics; its roots reached to Antiquity, and it was widely accepted. However, it continued to inspire discussions until at least the seventeenth century. In the

[1]For a description and the contextualization of the role of mathematics within the Jesuit *ratio studiorum* and the hierarchical organization of knowledge in accordance with the Thomist reform of the Aristotelian tradition, see Baldini (1992b, 19–56). Although a very general analysis, see also Dear (1987).

[2]Bos (1980, 329) even suggested that not only did this way of structuring the mathematical discourse have its own rights, but also that it possibly had a superior epistemology to that of nowadays. As he stated, "the terminology [of *mathesis pura* and *mixta*] is indeed an appropriate one, better than the division into 'pure' and 'applied' now in use, which overlooks the dialectical nature of the use of mathematics and suggests that one either practices pure mathematics or takes a ready parcel of mathematics and applies it elsewhere."

L. Guzzardi, *Ruggiero Boscovich's Theory of Natural Philosophy*, Science Networks. Historical Studies 60, https://doi.org/10.1007/978-3-030-52093-9_1

Prolegomena to the mathematical disciplines (*In disciplinis mathematicarum prolegomena*), which opened his *Commentary to Euclid's Elements*, Christoph Clavius appealed to the Pythagorean quadripartition of mathematics in arithmetic, music, geometry, and astronomy. Mathematics—Clavius explained, observing the Aristotelian canons of the Order—studies quantity, but quantity may occur as discrete (the number) or continuous (the magnitude).

> Arithmetic deals with discrete quantity as it is in itself [*secundum se*], inquiring and accurately explaining all properties and affections of numbers. Music treats that same discrete quantity or number as it is compared with another subject, in so far as it concerns the agreement and harmony of sounds; geometry also discusses magnitude (that is continuous quantity) in itself, in so far as it is immobile. Astronomy, finally, considers that same magnitude as far as it is mobile, such as the celestial bodies as far as they are moved of continuous motion. In fact, to those four mathematical sciences—of which arithmetic and geometry are called pure, whilst music and astronomy are called mixed—the other ones however dealing with quantity (that is, optics [*perspectiva*], geography and other similar disciplines) can very easily be reduced, as for the principles from which they shall depend. (Clavius 1611a, 3)

Clavius (*ibid.,* 3–4) also referred to another partition, which he ascribed to Proclus. According to the latter, he stated, "some of the mathematical science deal with intellectual things only, which are separated from any matter, whilst some other deal with sensible things, so that they have to do with matter submitted to the senses [*materiam sensibus obnoxiam*]." Arithmetic and geometry pertain to the first species; the second one includes astronomy, perspective, geodesy, music, practical arithmetic (called *supputatrix*, or the art of calculating) and mechanics, i.e., the science of machines.

Of course, this view could appear highly orthodox. After all, it was deep-rooted in the Aristotelian tradition. Therefore, most of Clavius' confreres, though not necessarily the mathematicians, could have been willing to share it without hesitation. However, Clavius used the pure-mixed partition instrumentally. When he described the "nobleness and excellence" [*nobilitas atque praestantia*] of mathematics, the reference to Proclus revealed a more distinct character:

> Since the mathematical disciplines deal with things that are considered apart from all sensible matter, although they themselves are immersed [*immersae*] in such matter, it is evident that they achieve an intermediate place [*medium locum*] between metaphysics and natural science, if we consider their subject, as Proclus correctly proves. The subject of metaphysics, indeed, is disjoined [*seiunctum*] from all matter in reality as well as in concept [*& re & ratione*]; the subject of physics, instead, is conjoined [*coniunctum*] to sensible matter in reality as well as in concept. From that, since the subject of the mathematical disciplines is considered without any reference to matter, although matter—according to the thing itself—is found in it, it holds clear that this is in the middle between the other two. (*ibid.*, 5)

From such a premise, Clavius inferred that mathematics was "not only useful, but also necessary." This, however, was not as undisputed as the pure-mixed distinction itself. Jesuits commonly shared the view that mathematics derived the certainty of its demonstrative procedures from the fact that theorems were reducible to chains of syllogisms. Clavius himself had asserted, and, to a certain extent, even practiced, the

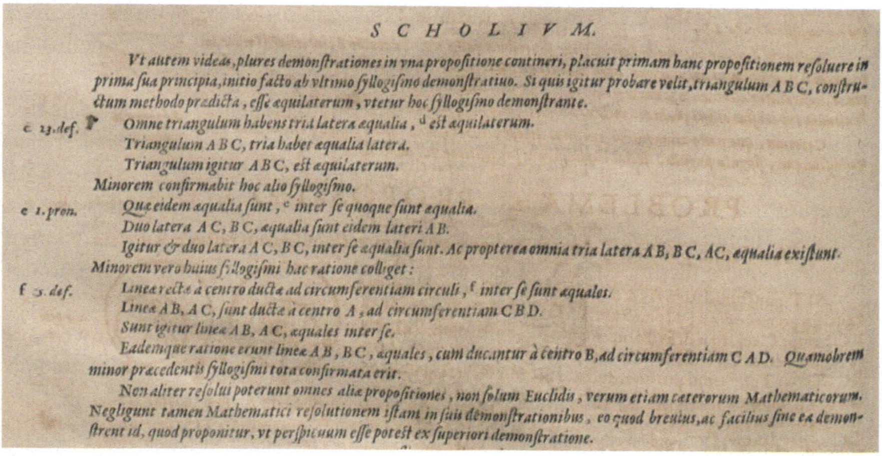

SCHOLIVM.

Vt autem videas, plures demonstrationes in vna propositione contineri, placuit primam hanc propositionem resoluere in prima sua principia, initio facto ab vltimo syllogismo demonstratiuo. Si quis igitur probare velit, triangulum A B C, constructum methodo prædicta, esse æquilaterum, vtetur hoc syllogismo demonstrante.

c 23.def. *Omne triangulum habens tria latera æqualia, [d] est æquilaterum.*
Triangulum A B C, tria habet æqualia latera.
Triangulum igitur A B C, est æquilaterum.
Minorem confirmabit hoc alio syllogismo.

e 1.pron. *Quæ eidem æqualia sunt, [e] inter se quoque sunt æqualia.*
Duo latera A C, B C, æqualia sunt eidem lateri A B.
Igitur & duo latera A C, B C, inter se æqualia sunt. Ac propterea omnia tria latera A B, B C, AC, æqualia existunt.
Minorem vero huius syllogismi hac ratione colliget:

f 15.def. *Lineæ rectæ à centro ductæ ad circumferentiam circuli, [f] inter se sunt æquales.*
Lineæ A B, A C, sunt ductæ à centro A, ad circumferentiam C B D.
Sunt igitur lineæ A B, A C, æquales inter se.
Eademque ratione erunt lineæ A B, B C, æquales, cum ducantur à centro B, ad circumferentiam C A D. Quamobrem minor præcedentis syllogismi tota confirmata erit.

Non aliter resolui poterunt omnes aliæ propositiones, non solum Euclidis, verum etiam cæterorum Mathematicorum. Negligunt tamen Mathematici resolutionem istam in suis demonstrationibus, eo quod breuius, ac facilius sine ea demonstrent id, quod proponitur, vt perspicuum esse potest ex superiori demonstratione.

Fig. 1.1 Clavius' syllogistic reduction of Euclid's first proposition ("Given a terminated straight line, to construct an equilateral triangle") as given in the scholium. "In order that you see that many demonstrations are included in a single proposition, it seemed proper to resolve that first proposition in its first principles, beginning with the last demonstrative syllogism. If anyone wants to prove that the triangle ABC, constructed by the method expounded, is equilateral, the following demonstrative syllogism shall be used." Then, a chain of three syllogisms is exposed; finally, as Clavius commented, "in no other manner can all other propositions be resolved, not only those of Euclid, but those of all other mathematicians as well. Nevertheless, mathematicians ignore such resolution in its demonstrations, because they demonstrate more briefly and more easily what they aim at, as can be clear from the demonstration given above"

equivalence between mathematical and syllogistic demonstrations. In the *Commentarii* to the *Elements*, he divided Euclid's first proposition ("Given a terminated straight line, to construct an equilateral triangle") into a chain of three syllogisms (see Fig. 1.1). He commented that "in no other manner can all other propositions be resolved, not only those of Euclid, but those of all other mathematicians as well." Nevertheless, according to Clavius, the syllogistic manner of presentation is not the favorite tool for most of his colleagues, "since they demonstrate in a shorter and easier manner what they aim at,"—i.e., in geometry at least, through a constructive procedure (*ibid.*, 28). Therefore, for Clavius, the syllogistic translation of a given theorem might be inconvenient, and usually it is, but in principle, it is always feasible.

However, mathematical syllogisms deal with hypothetical objects such as points, lines, triangles, and circles—even with physically problematic "incommensurable quantities." Physics, instead, deals with real objects: As Clavius himself recognized in the above-quoted passage, "the subject of physics is conjoined to sensible matter *& re & ratione*." Even if one was willing to admit that the ontological status of the mathematical entities does not affect the certainty of the demonstrations (and not all

Jesuits agreed on this),[3] a tension emerged: How could it be that, from premises related to hypothetical objects, one could draw real conclusions, i.e., conclusions related to physical, ontologically real entities?

To frame the subject of mathematics as an "intermediate place"—i.e., to state that mathematics is "immersed" in matter, though we can abstract from matter and consider the quantities with which it deals apart from it—was Clavius' strategy for answering that question. From this perspective, the objects of mathematics are not themselves hypothetical, but they are considered apart from matter. An extension of the mathematical methods to the physical sciences is as natural as a re-immersion of mathematics into the context from which it was abstracted.

This view perhaps best emerges in Clavius' treatment of astronomy. As we have seen above, Clavius accepted that it was more a *mathesis mixta* than an integral part of physics. This implied that an astronomical inference starts from premises that are partly mathematical (hence, hypothetical) and partly physical (hence, real). On the other hand, in a Thomist approach following *Posterior Analytics* (an obvious reference for the Jesuits), mixing experience and hypotheses prevented the observation of the causal order of the demonstrations themselves as well. In other words, the demonstrated features of a certain *real* entity (e.g., a planet) could not be the effect of the *hypothetical* features assumed at the beginning of the process (e.g., the mathematical assumptions about its position and state of motion).[4] Therefore, although astronomy was sound as a mathematical discipline, it could not provide

[3]See Feldhay (1987) and Baldini (1992b, 49–52). The fiercest opposition to Clavius' view is usually ascribed to Benito Pereira; as Feldhay (1987, 198) summarized in regard to his position, "he argued that the abstract nature of mathematical entities prevented the mathematical sciences from becoming real sciences, and that the mathematical disciplines do not have true demonstrations." Of course, the debates on the ontological status of mathematical entities and the scientificity of mathematics were the two most important issues of the *Quaestio de certitudine mathematicarum*. The related bibliography is substantial. A brief historical outline is sketched in Feldhay (1998, 135, n. 9). For a summary introduction, see Schüling (1969, 41–56) and Jardine (1988, 693–697). More detailed reconstructions are found in Giacobbe (1972a, 1972b, 1973), de Pace (1993), and Mancosu (1996, 10–33). In regard to the contributions of the Jesuits to the *Quaestio*, see, in particular, Giacobbe (1976, 1977), Feldhay (1998), and Baldini (1992b, 49–52; 1998, 710).

[4]See Feldhay (1987, 203–204) and Baldini (1992b, 42). This is the text of Thomas Aquinas in his *Comments to the Posterior Analytics* (lb1 lc25 n.6): "Then (79a13) he shows how *quia* and *propter quid* differ among sciences that are diverse but not subalternate. And he says that many sciences which are not subalternate are nevertheless related, i.e., in such a way that one states the *quia* and the other the *propter quid*. This is true of medicine and geometry. For the subject of medicine is not subsumed under the subject of geometry as the subject of optics [*perspectiva*] is. Nevertheless, the principles of geometry are applicable to certain conclusions reached in medicine: for example, it belongs to the man of medicine who observes it to know *quia* that circular wounds heal rather slowly; but to know the *propter quid* belongs to the geometer, whose business it is to know that a circle is a figure without corners. Hence the edges of a circular wound are not close enough to each other to allow them to be easily joined. It should also be noted that this difference of *quia* and *propter quid* between sciences that are diverse is contained under one of the modes previously discussed, namely, when the demonstration is made through a remote cause." (Translation by Fabian R. Larcher)

any physical explanation and was aimed at saving the phenomena through the use of geometrical tools *ex suppositione*.

Nevertheless, Clavius seemed to oscillate between this kind of traditional instrumentalist approach and a more daring concept of astronomy as a true *scientia* (in the Aristotelian sense), thus providing indisputable demonstrations and tracing back from the observed effects to the real causes. In Chapter IV of his *Commentary to the Sphere*, first published in 1570,[5] he dealt with, among other elements, "the circles and motions of the planets" and presented the theory of eccentrics and epicycles. Here, both evidence (*apparentia*) and arguments (*rationes*) for the use of eccentrics and epicycles in astronomy are displayed; in particular, Clavius' third and last argument is a methodological one:

> *Just as in natural philosophy we may arrive at knowledge of causes by their effects, so it is indeed in astronomy*, in which what happens to the celestial bodies goes on very far from us. It is unavoidable [*necesse est*] that we come to knowledge of them, of their disposition, and of their composition through their effects, that is, from the motions of the stars perceived through our senses ... Therefore when eccentric orbs and epicycles are such that by them astronomers can account for all phenomena easily ..., and when nothing follows from them that is absurd or incorrect in natural philosophy, as will soon be established by the solutions of the arguments that the adversaries of these kinds of orbs are wont to bring up, then have astronomers rightly declared that planets are conveyed in eccentric orbs and epicycles. (Clavius 1611b, 300; italics added)

The argument itself reduces to the idea that astronomical explanations by eccentrics and epicycles save the phenomena better than explanations by other means (i.e., homocentric spheres and the Copernican theory).[6] However, Clavius, in this case, also raised the important but controversial point that astronomy attains causal knowledge in the same manner as natural philosophy, which pertains to the physical reality: In regard to their mode of demonstration, they have the same level of scientificity. He was aware of his confreres' typical objection to this:

> The adversaries [to this view] attempt at weakening [*enervare*] this last argument by conceding that, once eccentric orbs and epicycles are posited, all phenomena can be defended; however, they say, from this does not follow that such orbs are found in nature [*in rerum natura reperiri*], they are instead entirely fictitious ... On the one hand, perhaps, all the appearances can be defended in a more convenient way (although it might so far be unknown to us); on the other hand, it may be that such orbs do indeed account for the appearances, albeit they are entirely fictitious and by no means the true cause of those appearances, just as from a false statement one may infer a true one, as follows from Aristotle's Logic [*Dialectica*]. (*ibid.*)

Clavius responded with a double-layered argument. The first layer leaned toward instrumentalism or, more precisely, toward a kind of *commodisme*, apparently

[5]Clavius' *Sphere* had a notoriously complicated publication history. For a reconstruction in the context of *The Tracts on the Sphere*, see Valleriani (2017, 451–454, 469–470).

[6]See Lattis (1994, 116–137). The subsequent quotations are also reported in Lattis; the translation given here may differ in some ways from his own—in all but one case, as I will explain, this is due to reasons of stylistics or uniformity.

without ontological commitment concerning the real existence of the theoretical-mathematical tools employed:

> First, if they have a more suitable [*commodiore*] way, they should show it to us, and we will be satisfied and extremely thankful to them. Indeed, astronomers demand no other than to consider all celestial phenomena in the most suitable manner [*quam commodissime*], be it done by eccentric orbs and epicycles or in another way. And since no more suitable [*commodior*] way has yet been discovered than that which defends everything by eccentrics and epicycles, it is quite believable that the celestial spheres are made of orbs of this kind. (*ibid.*)

However, in his response, he also introduced a new element, which unveils a different, *causal* layer:

> Thus, if they cannot show us a more suitable [*commodiorem*] way, surely they ought to assent to this way, which is inferred [*collectae*] from such a variety of phenomena: [in particular,] if they not only aren't willing to destroy utterly the natural philosophy, which is taught [*praelegitur*: in the sense of the German *vorgelesen ist*] in the schools, but also [do not want] to hinder the access to all the other arts, which inquire into causes by means of effects. Whenever someone adduces some cause from evident effects, I will state absolutely that doubtless another cause unknown to us could perhaps produce those effects. But for sure, if one has to assent to that cause being found, because it has some connection with the effects from which it is inferred [*collecta est*], one has also to concede eccentrics and epicycles, which have so much connection with the appearances that everything can be easily defended through their motions. (*ibid.*)

Astronomy, Clavius contended, does not reduce to a search for the instrumentally-more-convenient way to describe physical phenomena. It ought to save the phenomena, but this operation is grounded in a causal connection. The theory of eccentrics and epicycles, Clavius argued, is not sound simply because it saves the phenomena geometrically; rather, it is because it displays a causal connection among the appearances. Note that he partly agreed with the logical argument of his supposed adversaries: it is possible that another cause that is not yet known may produce the same effects, so other explanations are *feasible*. However, this simply asserts the fact of the underdetermination of any theory by the evidence at one's disposal, but it does not suggest anything about those feasible explanations. In other words, this is where logic should stop and leave room for "all the other arts, which inquire into causes by means of effects." For Clavius, astronomy was an inquiry of this kind.

Hence, Crombie (1975, 166) was not exaggerating when he stated, "Clavius gave a brilliantly lucid exposition of the criteria for deciding whether or not the spheres and epicycles, postulated in astronomical theory to account for the observations, had any real physical existence." Nevertheless, this does not mean that he was somehow defending the idea that mathematics is the real essence or nature or, to use Galilei's image, that God gave to us a book—i.e., the Universe—written in mathematical jargon. Clavius was not Galilei, and nature, for him, was not a book. Nature is nature; mathematics is mathematics. It is ours—not God's. But mathematics (astronomy, in this case) can grasp the real nature of things by launching hypotheses that experience proves to be the most appropriate.[7] To recognize that astronomy is a *mathesis mixta* is particularly relevant for this context; this means that eccentric orbs and epicycles are not merely mathematical tools or hypotheses. Instead, such tools stand for physical objects that are mathematically inferred and in which mathematics is immersed.

The reason for the power of mathematics is that it is capable of putting its ability to construct valid inferences at the service of other "arts." For example, in astronomy, and thus a case of mixed mathematics, one can start with observations and argue *mathematically* about the causes of some observed effect. As Clavius (1611b, 3–4) stated in the preface to the *Commentary to the Sphere*: "If we consider the mode of demonstration which is used in astronomy, nobody will deny that all the natural disciplines are surpassed by far by this science. And indeed, in order to confirm the things with which it deals, it employs very powerful [*efficacissimas*] demonstrations, namely the geometrical and the arithmetical, which, according to the opinion of all philosophers, have the highest degree of certainty."[8]

This, however, should not lead one to overlook the fact that, according to Clavius, there is a difference between astronomy as a *mathematical* discipline and the physical viz. natural disciplines—a difference that is implied in expressions such as "Just as in natural philosophy . . ., so it is indeed in astronomy" and "If we consider the mode of demonstration . . . all the natural disciplines are surpassed by far by [astronomy]." This also emerges from Clavius' often-quoted complaint (supposedly conveyed in 1582) about the low consideration that the official *curriculum studiorum* in the Jesuit schools directed toward the mathematical disciplines:

> First, it will be necessary that the students of physics classes at the same time have lectures in mathematical disciplines . . . For the experts agree that physics cannot be rightly understood without them, most of all as to that part which deals with the number and motion of celestial orbs, the multitude of angels [*intelligentiarum*], the effects of the stars (which depend on various conjunctions, oppositions, and other distances between them), the division of continuous quantities into infinities, the tides, the winds, comets, the rainbow, the halo, and other meteorological matters; the proportion of motions, qualities, actions, passions,

[7] I will not discuss the other important issue with which Clavius (1611b, 301) reinforced his theses, i.e., the predictive power of astronomy, which is suggested, e.g., by the following passage: "From eccentric orbs and epicycles, not only are the appearances of past things already known defended, but also future things are predicted, the time of which is completely unknown. So if I were to doubt whether, for example, at the full moon of September 1587 there will be a lunar eclipse, I can be sure from the motions of the eccentric orbs and epicycles that there will be an eclipse, so I would doubt no more." For a discussion, see Lattis (1994, 134–135).

[8] The asserted superiority of astronomy over the other physical sciences could have been irritating enough for Clavius' Jesuit colleagues, who possibly taught such disciplines. Notwithstanding, and quite ironically, he went on in an even more irritating manner: "Therefore, not without reason from both principles—of course, the nobleness of the subject and the certainty of the demonstrative process—Ptolemy holds, at the beginning of the *Almagest*, that astronomy simply is the first one amongst all the other sciences. He says that natural philosophy and metaphysics, if we look at their mode of demonstration, can be called conjectures rather than sciences because of the multitude and discordance of opinions [*appellandas potius esse coniecturas, quam scientias, propter multitudinem, & discrepantiam opinionum*]."

> reactions, etc., about which the Calculators [*calculatores*] wrote much. I omit countless examples in Aristotle, Plato, and their major commentators, which can by no means be understood without an appreciable knowledge of the mathematical sciences . . . It would be very helpful if the instructors of philosophy would abstain from those questions that can hardly be useful to understand nature [*parum iuvant ad res naturales intelligendas*] and lower very much the authority of the mathematical disciplines among students, for example those where they teach that the mathematical sciences are not sciences, do not have demonstrations, abstract from the being and the good, etc. (Clavius 1582, 116)

In the end, Clavius campaigned more for a higher consideration of the mathematical sciences than for replacing the traditional *philosophia naturalis* with them. What he sought was no less and no more than an orthodox extension of the mathematical methods to physics—an extension, however, that his colleagues in the Jesuit citadel of science were hardly willing to concede.

1.2 Light from Abroad: The Flandro-Belgian Connection and Gilles-François de Gottignies

Dear (1987, 165) contended that "Clavius' largely successful attempts to give mathematics a significant place in the Jesuit educational curriculum had created teachers and professors of mathematics in the colleges whose academic position in principle equaled that of the natural philosophers." This was certainly one of Clavius' best purposes; it is debatable as to whether he succeeded during his late years and when this change occurred. The fate of the Academy of Mathematics (Baldini 2003, 51) and the stories of his immediate collaborators and successors at the Roman College immediately after his death (1612) seem to prove that this aim required a considerably long timespan and was the effect of a plurality of actors under the long shadow of Clavius.

Although, in the 1590s, his increased prestige within and outside of the Society of Jesus ensured a better reception of his projects regarding an enhanced role of mathematics, after Clavius' death, for reasons that are not entirely clear, the mathematical tradition of the Roman College rapidly began to lose ground. The first signals had been encouraging, though: Since the mid-1590s, Clavius had relied on his closest pupils for the teaching of mathematics at the College, keeping for himself the organization and supervision of the "Academy of Mathematics". After its informal beginnings in the 1580s, in the decade that followed, it developed into a special pedagogical unit for the teaching of advanced mathematical courses. The standard courses were alternatively taught by a number of mathematicians, including both reliable students and collaborators ,of Calvius, such as Christophorus Grienberger (1595–98, 1602–5, 1612–16, 1624–25, 1628–33), Odo van Maelcote (1605–1610), and Orazio Grassi (1616–1624, 1626–1628), as well as other minor figures, such as Gaspare Alperio (1598–99, 1600–1602), Angelo Giustiniani (1599–1600), Vincenzo Filliucci (1610–11), and Niccolò Zucchi (1625–1626).[9]

Maelcote and Grienberger, in particular, ensured a high level of technical knowledge and noticeable openness to the new science in the mathematical school of the Roman College. Having completed his novitiate in Tournai and concluded his studies in Douai, in the Spanish Netherlands, in 1601, Maelcote was summoned by Clavius to Rome. Thereafter, Maelcote began to assist Clavius and Grienberger at the Academy and taught several courses of mathematics. In April 1611—together with Clavius, Grienberger, and Giovanni Paolo Lembo—he was asked to review Galilei's astronomical work. The following month, Maelcote's effort resulted in an address—in fact, a public praise of Galilei in front of the College—titled *Nuncius Sidereus Collegii Romani*.[10] However, Maelcote's career was abruptly halted due to his death in 1615. In the following years, Grienberger, who had also pressed for greater openness, like other confreres, found himself bound to the defense of the official scholastic Aristotelianism of the Society of Jesus. He also discovered that he was bound by the obligation to avoid codified prohibited opinions, particularly in cosmology.[11] With the Galileo Affaire in the background, the dictate applied to the Society of Jesus as a whole, but in Rome, the fulcrum of Jesuit science and power, the pressure was obviously highest. As Baldini (2003, 68, 79–80) concluded, all of this led to a defensive attitude at the Roman College, which resulted in lesser attention to the technical skills of the new generations of mathematicians and made the prosecution of high-level research difficult along the lines that Clavius had drawn.

Other promising *mathematici* who formally or informally followed the Academy of Mathematics in Clavius' late years were sent abroad to their original provinces. This was the case for another student from Douai, who had come to Rome in 1606—namely, the Belgian Grégoire de Saint-Vincent, who was perhaps the most skilled among Clavius' pupils. Born in Bruges in 1584, he entered the local Jesuit college in 1595; starting in 1601, he studied philosophy and mathematics in Douai. Four or five years later, he was in the Jesuit novitiate of S. Andrea in Rome; after that, possibly based on Clavius' advice, he began to attend the courses of philosophy and theology and became acquainted with the mathematicians' circle, attending the special classes of the Academy from 1607 until Clavius' death (Baldini 2003, 79 n.25, 96 n.106).[12] In that year, he most likely returned to Belgium, where he became a priest in 1613. In the following period, he was assigned to teach Greek in different colleges of the

[9]Baldini (2003) provides an analytic reconstruction of the history of Clavius' Academy of Mathematics. For the list of the professors of mathematics after Clavius, see Villoslada (1954, 335); see also Lattis (1994, 24).

[10]On Maelcote, see Sommervogel (1904, V, col. 281–282), Villoslada (1954, 321, 335) and Vanpaemel (2003, 394–395). On his laudatory address to Galilei, see, in particular, Lattis (1994, 187–195) and Reeves (1997, 151–152).

[11]For Jesuit censorship in this context, see Feingold (2003, 18–22).

[12]Much of Saint-Vincent's early life is only approximately known. For a biography, see van Looy (1981).

Flandro-Belgian Province, and also served for a year as an almoner to the Spanish troops quartering in the Netherlands. In 1617, Grégoire was appointed as a professor of mathematics, succeeding François d'Aguilon in Antwerp; he then taught in Louvain in the 1620s and in Ghent from 1632 to his death in 1667.

Equipped with a Clavian background, Grégoire viewed d'Aguilon as a "brother in arms" (Vanpaemel 2003, 396), in so far as the latter's efforts in the previous decade had paralleled those of Clavius in Rome 20 years earlier. His *Six Books of Optics*, printed in Antwerp in 1613, featured the significant subheading of "useful to the philosophers as well as to the mathematicians" (*Philosophis iuxta ac mathematicis utiles*). Moreover, the introducing epistle "To the Reader" raised a paean to the importance of optics, "queen of all the sciences that mathematics embraces" (d'Aguilon 1613, "*Ad Lectorem*", s.n., initial page). It closed with a praise of the "mixture" of mathematics and matter, which is reminiscent of Clavius' discourse on the "intermediate place" of (mixed) mathematics between metaphysics and physics:

> I will hardly agree with Plato, who argued that nothing is mathematical that is not separate from matter [*a materia seiunctum*]: he also maintained, in contrast to Eudoxus and Archytas, that the mechanical art and the mode of demonstration through instruments do not deserve the name of mathematics. ... In this regard, it is more sound to refer to Archimedes, who believed that the conjunction of things with matter [*rerum cum materia copulationem*] in no ways detracted anything from mathematics, rather it enriched and brought to completion that science. Trusting in his authority, we have undertaken the cultivation of this part of mathematics that comprises both genres [*genus*], namely the philosophical and the mathematical. (d'Aguilon 1613, "*Ad Lectorem*", final page)[13]

The "philosophical kind" in the last quoted sentence reflects the Jesuit *ordo studiorum ac scientiarum*, which assigned the teaching of *physica* to the philosophers. As we have seen, this was one of the motives that inspired Clavius' 1582 complaint against the "instructors of philosophy"; in alignment with him, d'Aguilon advocated the use of mathematical methods in optics—not optics as a *mathesis mixta*, but physical optics, concerned with the process of vision (exposed in Book I), the "optic ray" and the horopter (Book II), visual cognition and the visual features of objects (Book III), fallacies in perception (Book IV), luminous and opaque bodies (Book V), and projections (Book VI). In all of the books, d'Aguilon structured his propositions in mathematical fashion as theorems, lemmas, and—more rarely—*consectaria* (i.e., corollaries). In certain instances, he also included definitions ("*praenotationes*", i.e., premises or precognitions), hypotheses, and axioms. Sometimes, this comes across as a rhetorical ploy, rather than a result of a theoretical need. In Book I, for example, the "theorems" are mostly reduced to descriptions of

[13]Here is the original text: "*Non enim facile Platoni calculum adiecero, qui nisi quod a materia seiunctum esset, Mathematicum censebat nihil: artemque machinalem, & organicum demonstrandi modum adversus Eudoxum & Architam Matheseos nomine indigna reputabat. ... Sanius hac de re Archimedes, qui rerum cum materia copulationem non modo nihil Mathesi derogare, sed vero exornare eam etiam ac perficere existimavit. Cuius nos auctoritate freti hanc Matheseos partem, quae utrumque genus, Philosophicum scilicet & Mathematicum comprehendit, suscepimus excolendam.*"

apparatuses and phenomena of vision. In other cases, in which the mathematical content prevails (as in Books II, IV, V, and VI), d'Aguilon applies geometric constructions and demonstrations in the strict sense, often concluding with typical expressions such as "*quod erat probandum*" (II, Prop. XXI, XXXIX; III, Prop. III; V, Prop. XVI), "*quod propositum fuit demonstrare*" (II, Prop. XXXIX; V, Prop. XX), "*quae omnia ex ... lemmate sunt manifesta*" (V, Prop. XXI), and "*quod demonstrare oportuit*" (V, Prop. XXXII). There are obviously many other examples from many other places (most of all from Book VI); however, an enumeration of such examples is not relevant to the scope and aims of this chapter.

It is not clear whether d'Aguilon was consciously echoing Clavius, as one might suspect based on expressions such as "*a materia seiunctum*" or "*rerum cum materia copulationem*" in key places of the above-quoted passage, or whether he had independently developed a similar epistemology. To be sure, Clavius' works circulated within the Jesuit colleges' network and above; as a matter of fact, d'Aguilon (1613, 434) quoted the *Commentary to the Sphere*, in which Clavius had expressed his epistemological concerns. A more exact reconstruction of the routes that these ideas might have taken would certainly be crucial in providing further evidence.[14] For the purposes of the present chapter, however, the sketched background should be sufficient in order to understand how and why the idea of an extension of the mathematical methods to physics, which had characterized much of Clavius' and d'Aguilon's epistemology, would *not* re-surface at the Roman College until the 1660s.

In 1661, Gilles-François de Gottignies became a professor of mathematics in the Jesuit citadel of science. He had arrived from the Flandro-Belgian province 1 or 2 years earlier in order to conclude his *curriculum studiorum* with the course of theology. Born in Brussels in 1630, Gottignies was a student of André Tacquet in Louvain during the late-1640s and the early 1650s. In 1653, he entered the Jesuit novitiate in Mechelen (Malines); some years later, the *Catalogi Personarum* of the Jesuit college of Ghent reported his name in the years 1657–1659, along with the qualification "*mathematicus*." Thus, he was a pupil of Grégoire de Saint-Vincent. Educated in the Belgian tradition, through Grégoire, Tacquet, and, indirectly, d'Aguilon, he may have absorbed the Clavian values of a high esteem of

[14]Another association with Clavius is d'Aguilon's activity as the organizer of mathematical instruction in the Flandro-Belgian province. Shortly before his death, he promoted and established the school of mathematics at Antwerp, where Grégoire of Saint-Vincent was to teach. Officially, the school was initiated in order to serve the interest of the Academy of Church History (a newly-founded institution chiefly devoted to Counter-Reformation apologetics) on matters of chronology. Arguably, though, d'Aguilon put into this effort all of the motivations for mathematical teaching that he had expressed in his *Six Books of Optics*. Unfortunately, the details of the school's early programs are not preserved, and d'Aguilon died in 1617, before starting the official courses: see van de Vyver (1980, 265–266). Nevertheless, the institutional aspect might be relevant to the transmission of knowledge and epistemology from Clavius to Gottignies through Grégoire of Saint-Vincent and François d'Aguilon.

mathematics and its relative autonomy that, after Clavius, had largely been overlooked in Rome.[15]

A discussion of the relationship between mathematics and physical science leads to Gottignies' undated manuscript titled *Pilae motae et quiescentis considerationes Physico-mathematicae*. It is possibly from a course of lecture that according to Baldini (1992a, 38) should be dated between 1667 and 1684, a date after which Gottignies ceased teaching:

> The physico-mathematical sciences . . . have their origin partly in pure physics, partly in pure mathematics. . . . Physico-mathematical sciences are for example, optics, statics, geography, astronomy and others . . . I state first: of course pure physics, that is to deduce and establish conclusions from physical principles, properly concerns only physicists and not mathematicians; and in turn to deduce conclusions from pure mathematical principles properly concerns mathematicians, not physicists. For in any science, what is characteristic to that science concerns those who profess that science. I state second: to deduce conclusions partly from physical, partly from mathematical truths properly concerns to mathematicians, not to physicists. Namely, those conclusions are on their own physico-mathematical conclusions . . .; to establish them properly concerns those who profess physico-mathematical sciences, but such are only mathematicians. (Gottignies [1667], 4v–5v.)[16]

Similar to Clavius some decades earlier, Gottignies did not view mathematics as necessary to the grounding of physical knowledge. This would have placed him outside of the limits of the Jesuit teachings, because it would have implied the reality-in-nature of the mathematical entities. Rather, he carefully distinguished between physical and mathematical competences. Mathematicians and physicists do not consider the same subject, but superposition—hence, collaboration—is possible. The resulting subject is the *physico-mathesis*.

---

[15]Scholarship has chiefly concentrated on his *logistica*, "an attempt at providing better foundations for algebra," as Leibniz described it in 1702 in a letter to Varignon (see Mancosu 1996, 89–90). Sommervogel's (1890, 3, 1624–1626) work on him is highly inaccurate and problematic; more precise biographical notes are to be found in Siret (1884) and Mols (1986).

[16]The original Latin text (in a more extended form) reads as follows: "*. . . Verum etiam maxime necessari[a]e erunt ex Physica-Mathesi coniuncta, nimirum ill[a]e omnes quae Physico-Mathematicas appellant, quia origine habent partim ex pura physica partim ex Mathesi pura, et quoniam eiusmodi scientias dari certum est, negari non potest, quod si ad physicum spectet scientias phyisico-mathematicas tractare aliquod illi ius esse in mathesim, et similis si Mathematici optimum est Physico-mathematica tractare, ipsi aliquod ius competere in Physicas; Scientiae Physico-mathematicae sunt ex. gr. Optica, Statica, Geographia, Astronomia et alia similes in quibus quid quid illis proprium inuerit, id neque ad puram physicam, neque ad puram Mathesim spectat, sed ex iis eruit, quae predictis duabus scientiis propria sunt; cur autem scientiam physico-mathematicam tractationem non sibi vindicent qui vulgo Physici aut Phylosophi [sic] appellant sed illitantum, qui profitent Mathesim, et Mathematici vocant, parum refert . . . | [5r] Assero primo. Sane pura Physicas hoc est ex solis principiiis physicis conclusiones deducere atque stabilire proprie | [5v] ad Physicos aspectat, non ad Mathematicos et vicissim conclusiones ex puris principiis mathematicis deducere proprie ad Mathematicos pertinet, non ad Physicos, quod enim cuiquae scientia proprium est ad illum pertinet qui talem scientiam profitet. Assero 2°. Conclusiones partim ex Physicis, partim ex Mathematicis veritatibus deducere, proprie spectare ad Mathematicos non ad Physicos. Et enim conclusiones illae propriae sunt conclusiones Physico-mathematicae, . . . eas stabilire proprie spectat ad eos qui Physico-mathematicas profitent sed tales sunt soli Mathematici. Etca.*"

Of course, this label and cognate expressions were frequently used when Gottignies wrote his *considerationes* and had a variety of connotations. Dear (1995, 168–179) traced its first appearance back to a marginal gloss regarding Descartes that the Dutch mechanical philosopher Isaac Beeckman had reported in his journal from late 1618; here, it means a rather generic connection between mathematics and physics. By the middle of the seventeenth century, however, the attribute "physico-mathematical" became more common in both vernacular and Latin texts. Meanwhile, its use became more specialized. At times, it was attached to the pretensions toward the "new" mathematical approach and the "outdated" Aristotelian physics. During Gottignies' years as *professor matheseos* at the College, for example, a so-called *Academia fisicomatematica* was established and flourished in Rome. In this case, the label referred to the "natural philosophy based on experiment, in imitation of that celebrated Academy del Cimento ... and of other well-recognized ones in England, France, and Germany," as described by Girolamo Tiraboschi in his comments on some of the minutes of the Academy's Secretary (Knowles Middleton 1975, 143–144).

However, Gottignies seemed to refer to a more traditional distinction that was crucial in the Jesuit order of sciences: namely, the distinction between a "subalternating" and a "subalternated" hierarchic level of the sciences.[17] This differentiation was in operation, for example, in Riccioli's *Almagestum novum* (1651, I, 2), in which astronomy was described as "a physico-mathematical science of the terminated quantity of the celestial bodies and of the terminating quantity of their sensible accidents." Riccioli immediately made it clear that, as such, astronomy is "subalternated to physics" in regard to the object (the skies, as well as the celestial bodies and their features, such as figure, color, light, positions, motions). However, he also stated that "most of all it is subalternated to mathematics," for it considers all of this as a terminated quantity.

This all might appear to be very traditional. Still, something important had changed from Clavius' time. He had considered "optics, statics, geography, [and] astronomy" as mixed-mathematical disciplines: it was no wonder that the mathematicians, and not the physicists, were entitled to deal with them. Moreover, Clavius only focused on ensuring that the mathematical disciplines were on par with the philosophical disciplines. Riccioli and Gottignies, however, changed the status of these disciplines. They—especially Gottignies—labelled them as *physico-mathematical* sciences, which implied a tighter connection between mathematics and physics. But the scholars entitled to deal with these disciplines, as both Riccioli and Gottignies maintained, are still (and exclusively) the mathematicians.

Many of Gottignies' confreres might have considered his attempt to extend the mathematical methods to physics as an indication of the mathematicians' expansionism. However, his proposal remained within the hinges of Jesuit orthodoxy,

[17]See Dear (1995, 168–169). For the medieval roots of this distinction and its relationship with Aristotle's *Posterior Analytics*, see Wallace (1972, 1: 30–38, 77–80). For its relevance to the Jesuit hierarchic system of sciences, see Baldini (1992b, 20–73).

taking for granted that mathematics follows an efficient "syllogistic" demonstrative method and that pure mathematical entities might be hypothetical. The world is not mathematical in and of itself, but mathematics is immersed in nature as *quantitas terminata*, subject to measurements and studies through mathematical methods. Thus, such methods can be used in physics, and this use can be labeled as *physico-mathesis*. What is important in this approach is the process through which results are achieved, rather than the ontological status of the individual mathematical entities involved. Moreover, the fact that such a process is mathematical ranks it among the tasks of the mathematicians: only they are entitled "to deduce conclusions partly from physical, partly from mathematical truths." Mathematicians have to grant that the truth of the premises is preserved in the chain of derivations. However, the physical truths, particularly in astronomy, are given in advance, and mathematics is neutral about them. A discussion of their truthfulness remained above the competences of the mathematicians, but half a century after Clavius' death, they attained the freedom to do what they wanted within boundaries that had expanded considerably.

1.3 Desperate Defenses in a Physico-Mathematical Style

In the following paragraphs, I shall employ the phrase *physico-mathematical style* to mean the approach that traces back to Gottignies. Of course, I do not claim that this was the essence of his activity or represented its most important part. Rather, I limit myself to claiming that Gottignies, because of the importance of his role as *professor matheseos* at the Roman College, can be viewed as a prominent representative of a widespread tendency in the practice of mathematics within the Society of Jesus and elsewhere. In brief, this style consists of instrumentally representing magnitudes occurring in nature through geometrical or algebraic magnitudes and implies that the mathematical conclusions, which are necessarily correct because of the mathematical method employed, should represent the physical consequences. It is a style in so far as it consists more of a way of treating an object and presenting results rather than a discourse about the meaning and ontological status of the entities involved.[18]

An important corollary of the physico-mathematical style, derived from the tradition of Clavius, is that, if the physical effects do not match the mathematical conclusions, the *physical* theory is mistaken and should be modified or recused. In particular, the physico-mathematical method could not be employed in order to explain the ultimate cause of motions, which is a metaphysical or possibly a theological task. But it could certainly be applied for testing divergent explanations

[18]By using the concept of style, I also intend to hint at Hacking's (1992 and 2009) idea of style of reasoning or scientific style, as well as at Granger's (1968) concept of mathematical style. In particular, the physico-mathematical style can be viewed as a radicalization of Hacking's (and Crombie's) "Galileian" style.

of why certain observed motions take place. For this reason, such an approach could be put to good use to serve one important duty of a *Jesuit* mathematician: the defense of the orthodoxy in an area as crucial as cosmology.

For causes that are not the aim of the present book to discuss, the obligation to defend the geostatic theory lasted until 1757, when the so-called "anti-Copernican decree" of March 5, 1616, was upended (see Casini 1983, 143–155; Baldini 1992a, 29–30). Of course, this was somewhat of a "physical truth"; as such, the mathematicians could not dispute it. However, since astronomy was now perceived as a physico-mathematical science, they—and particularly those at the Roman College—were expected to use mathematics in order to defend the *hypothesis terrae quiescientis* by disproving the opposite view. As is well known, the 1616 decree banned "the false Pythagorean doctrine, altogether contrary to the Holy Scripture, that the Earth moves and the Sun is motionless," and a 1619 update clarified that the prohibition also applied to "all books that teach the motion of the earth and the immobility of the Sun."[19] Yet, it did not prescribe which system should be followed, thus allowing room for a plurality of options, which ranged from some versions of the Eudoxian homocentric spheres to Ptolemy's eccentrics and epicycles to the Tychonic "combined" system. However, in the second half of the seventeenth century, while the acceptance of Copernicanism was becoming widespread outside of the Society, homocentrics was no longer an issue, whereas the Tychonic alternative still provoked much interest at the Roman College.[20]

According to Tycho Brahe, both the sun and the earth were centers of motions, but they did not have the same status. As in Ptolemy's scheme, the earth was central and stationary, while the sun (and obviously the moon) orbited it; in turn, the planets orbited the sun, such as in Copernican cosmology. The Jesuits may have been sympathetic toward this option: based as it was on the earth at rest, ideally placed in the center of the cosmos, it could satisfy the Decree's requirements and neutralize the objections raised by the theologists of the Order. Its physical implications were also extremely flexible. It not only saved evidence that was hardly compatible with a Ptolemaic scheme (famously, this is the case of the phases of Venus), but also

---

[19] Antonio Favaro published the original text of the 1616 anti-Copernican decree in the documents complementing Galilei's collected works (see *Le Opere di Galileo Galilei*, Edizione Nazionale, ed. by A. Favaro, Firenze: Barbera 1938, vol. 19: 322–323). For the English translation quoted here, see Finocchiaro (1989, 148–149). Other details about the decree and its 1919 update are provided in Mayaud (1997, 48–52) and Finocchiaro (2005, 16–20).

[20] In the last decades of the sixteenth century, homocentrics appeared as a valuable rival hypothesis to Clavius. In his 1570 *Commentary to the Sphere*, he argued for the Ptolemaic system and attempted to offer some good mathematical arrows for its bow; he also discussed (unfavorably) the rival hypotheses, chiefly some versions of the homocentrics and the Copernican theory. (A more elaborate criticism of Copernicus appeared in the revised 1581 edition.) However, Clavius did not include a review of Tycho's system, which he must have known and probably considered mistaken in many ways (as Lattis 1994, 205–208, convincingly shows based on documentary evidence from Clavius' correspondence). This is quite striking when one considers that Tychonic-like systems became the favored cosmological options, particularly at the Roman College after Clavius (Lattis 1994, 208–211).

allowed for the removal of the Ptolemaic idea, still supported by Clavius and others, of rigid celestial spheres in which planets moved. Moreover, it was compatible with a "fluid-heaven" cosmology, which found an increasing number of followers within the Order (Lattis 1994, 211–216). With these premises, generations of Jesuit astronomers after Clavius became fascinated with Tycho's alternative.

Finally, Tycho's suggestion preserved much of Copernicus' architecture, hence its appeal to the mathematicians. They only had to show that the mathematical implications of Copernicus' idea of a moving earth did not match with the observations, whereas some variation of the Tychonic scheme saved the phenomena. Therefore, for the "physico-mathematicians", Tycho could be the perfect picklock to break the Copernicans' defenses and, in turn, offered the opportunity to secure the *hypothesis terrae quiescentis*. Its fortune lasted beyond the end of the seventeenth century, chiefly because of the mixed strategy that it allowed: refute Copernicus and simultaneously defend some variety of a stationary earth theory.

In the late seventeenth century, Gottignies' immediate successor, Francesco Eschinardi (who covered the professorship for only 2 years, from 1684 to 1685), pursued this approach; in the first decades of the following century, its most radical and mathematically skilled supporter became Orazio Borgondio, one of Gottignies' pupils and Boscovich's mentor.[21] As the curator of the Kircherian Museum, which he managed to equip with new astronomical instruments, Borgondio was a skilled astronomer. According to an early biographer, he "became very famous because of his astronomical observations" (Mazzuchelli 1762, p.1771). He also credited him with a work on "the system of Descartes" that reached the Acadèmie Royal in 1730; both Sommervogel (1890, I, col. 1807) and Villoslada (1954, 238) reported the information, but there is no evidence of such a work in the *Histoire* and *Mémoires de l'Académie Royale des Sciences*. He most likely had knowledge of Newton, but his dissertations do not mention him or any technical aspects related to the *Principia*.[22]

His own works and the dissertations defended by his students hardly convey the impression that Borgondio was an original researcher, preoccupied as he was with presenting established solutions rather than novel insights. As was typical practice in the Jesuit Colleges and beyond, he authored most of the dissertations defended by the pupils;[23] among others, two refutations of the Copernican theory are ascribed to him: *De telluris motu in orbe annuo ex novis observationibus impugnato theses mathematicae* (1714) and *De situ telluris exercitatio geographica* (1725). A third

[21]Baldini (1992a, 61–62) lists the professors of mathematics at the Roman College, together with the corresponding years of activity, between Gottignies (1661–1684) and Boscovich (1740/1741–1763): Francesco Eschinardi (1684–1685), Francesco Antonio Febei (1685–1686), Antonio Baldigiani (1686–1707), Giovan Francesco Musarra (1707–1708), Domenico Maria Turani (1708–1710), Ignazio Guarini (1710–1712), and Orazio Borgondio (1712–1740). Concerning Eschinardi's Tychonic-based defense of the *hypothesis terrae quiescentis*, see Beretta (2008, 550–551).

[22]For a biography of Borgondio, see Casini (1970). Further bibliographical information is given in Baldini (1992a, 29–30 and 62 n.43).

[23]On this kind of "transferred authorship", see Weijers (2013, 218–219).

dissertation directly relating to astronomy is significantly titled *Hypothesis planetarum elliptica Exercitationis Astronomicae gratia explicanda* (1732). Whereas the last astronomical treatise does not aim to recuse a particular theory but is an attempt to expose a Tychonic-like system with planets in elliptical motion,[24] the attempted refutations of 1714 and 1725 adopt a physico-mathematical style.

The background of Borgondio (1714) is the search of the annual parallax of the stars as a crucial test for or against the motion of the earth. A moving earth would imply that a periodic variation in the observed position of the stars should occur within a period of 1 year; so, it was quite obvious that the search for the annual stellar parallax could provide strong support for Copernicus' theory. Early efforts to detect a stellar parallax trace back to Galilei's time (Siebert 2005), but Robert Hooke is often viewed as the first Copernican to try to observe such a parallactic displacement by means of a telescope in order to prove the earth's motion, resulting in his *Attempt to Prove the Motion of the Earth from Observations* (London, 1674). Here, he gave an account of a 5-year long survey of γ Draconis (Eltanin), a star in the constellation of Draco that culminated near the zenith over Gresham College, where he was observing. However, many of Hooke's contemporaries considered his contentions rather puzzling. Albeit they mostly accepted his observations, they (even the Copernicans) generally doubted that these should necessarily be explained in terms of an annual parallax (Siebert 2007). So, to the delight of the anti-Copernican front, no consensus regarding such an effect was reached during the seventeenth century (and beyond).[25] And of course, a failed search of the parallax beyond a reasonable doubt could deliver a potential weapon into the hands of Copernicus' opponents.

Borgondio's dissertation *De telluris motu* (1714, 5) began by explicitly leaving aside theological arguments and instead developing a geometrical translation of an argument that traces back to Riccioli (1651, II, 452). At first, it follows the classical pattern of counterfactual reasoning: what if Copernicus was right? To answer the

---

[24]"The elliptical hypothesis places the planets, with their own motion driving them away [*motu proprio delato*], in the perimeter of an ellipse, whose kind is known. As a matter of fact, however much the lunar motion is varied so that the kind of ellipse is thought to change, at the syzygys, the axes of Lunar ellipse are in a ratio of 59–58.945; the Solar ellipse is 10–9.99985. In addition, the common focus of both ellipses is placed in the center of the earth. The ellipses of the inferior planets though, have their focus in the center of the sun in such a manner that the ratio of Mercury's axes is . . .". (Borgondio 1732, IV.) Note that planets move *motu proprio*, which signals an obvious difference with any Newtonian account, in which an external force moves planets (or because of the presence of another celestial body, e.g., the sun and the earth, *each of them* moving because of the presence of the other). To my knowledge, this is the most extensive and accurate application of Kepler's ellipses to the Tychonian System, although elliptical motions might have been considered by other antecessors of Borgondio. For example, Francesco Eschinardi (professor of mathematics at the Roman College in 1684–1685 after Gottignies) expounded a "*hypothesis elliptica*" in a Tychonian context in his work (1689, 117–118). However, here, he considered the elliptical motion as the effect of an epicycle (a result that was well known at the time: see Boyer 1947); moreover, contrary to Borgondio, there is no direct or indirect reference to Kepler's laws.

[25]On the history of the stellar parallax, see also Hoskin (1982, 29–36), Hetherington (1988, 16–21), and Hirshfeld (2001).

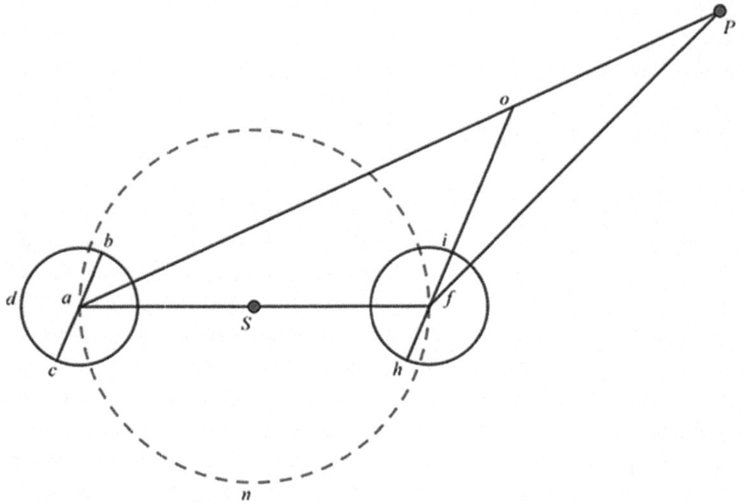

Fig. 1.2 Borgondio's diagram of the variation in the annual star parallax if Copernicus' theory is right. (Modern adaptation from Table I, Fig. 1, of Borgondio 1714. In the original diagram, the circle *bnf* is very flattened, similar to an oval or an oblong ellipse, but Borgondio 1714, 1, clearly states that "center *a* of the Earth *bdc* describe[s] every year, around the Sun *S*, the circumference of the circle *anf* [*telluris* bdc *centrum* a *decriberet quotannis circa solem S circumferentiam circuli* anf]")

question, a geometrical demonstration is combined with physical measurements. The result is a *reductio ad absurdum* according to this workflow: (1) assume Copernicans' point of view and reduce it to a geometrical scheme; (2) draw from this scheme physical-geometrical consequences, including the expected relative positions of the earth and the star(s); (3) finally, compare the geometrical conclusions with the data: if they mismatch, the theory is absurd.

So, let us assume the Copernicans' point of view and represent "the annual motion of the earth" through a diagram. In Fig. 1.2, *S* represents a fixed sun, around which the earth *bdc* orbits along the circle *anf* until reaching the position *igh*; the external point *p* represents a fixed star. Two other physico-mathematical assumptions should be made: (1) Let the semi-diameter *aS* of the circular orbit *anf* be the distance of the center of the earth from the center of the sun; and (2) let us assume that the terrestrial axis always preserves its inclination with respect to the orbital plane in which *anf* lies. (According to Borgondio, this is reasonable because "it properly explains the sequence of the seasons.")

On this basis, and looking at the diagram, one can demonstrate the following:

> When the center of the Earth is in *a*, a ray of any fixed star in *p* forms, together with the terrestrial axis *bc*, an angle with inclination *bap*; when the center of the Earth in *f*, a ray of a fixed star in *p* forms, together with the terrestrial axis *ih*, an angle with inclination *if*. Since the segments [*rectae*] *bc*, *ih* are supposed to be parallel [for the assumption 2], the ray *pa* meets the segment *bc* in *a*; moreover, as proved by Clavius in the scholium to Axiom 13, it will meet the segment *ih*, further prolonged, in any point *o*. But, from [Euclid's] *Elements*, Book I, Prop. 29, the angle *bap* is equal to the angle *aof*, and this, from [Euclid's] *Elements*, Book I, Prop. 16, is greater than the angle *ofp*; therefore, the angle *bap* is greater than the

angle *ofp*; hence, if the Earth moves, the fixed stars will change their inclination towards the terrestrial axis. (Borgondio 1714, 5)[26]

However, Borgondio offered this reminder to readers: “Neither Tycho nor Riccioli, employing many of their best instruments for that aim, observed such variation. Hence, the Earth does not move in annual orbit.” However, as the Copernican system increasingly gained consensus throughout the seventeenth century, the lack of a detectable annual stellar parallax turned this search into a tenacious hunt. Various measurements were performed during the second half of the century, though none of them were satisfying. The English astronomer John Flamsteed made a remarkable attempt; his results displayed an annual variation in the parallax of the Pole Star and were brought to the public in the form of a letter addressed to John Wallis (see Wallis 1699, 701–708).

Was it revenge for the Copernicans? If Borgondio wanted to prove that, notwithstanding the updated knowledge, the earth does not move, he needed to perform a more sophisticated construction than his first diagram, which referred to the state of knowledge in 1651, when Riccioli’s *Almagestum Novum* was issued. Therefore, he provided his dissertation with two other geometrical constructions with which he proved the following theorem (*Propositio III*): If we assume that the earth moves around the sun, “the inclinations of the Pole Star towards the terrestrial axis will be different about the equinoxial periods, and will be less about one of the equinoxes than in summer” (Borgondio 1714, 8). In astronomical terms, this means that the pole star declination (i.e., its elevation from the celestial equator expressed in degrees) should be higher on June 21 than during the vernal (March 21) or the autumnal equinox (September 21).[27] On the contrary, in a final corollary to the last proposition, Borgondio (1714, 10) revealed that Flamsteed’s observations did not agree with this prediction. In particular, for Flamsteed, the pole star would display no changed “inclination” toward the terrestrial axis during both equinoxes, while declination would be the lowest in June (or, in Borgondio’s terms, inclination would increase during the equinoxes and would decrease during the summer). Hence, Borgondio (1714, 10) concluded, “as a consequence, the Earth does not move in annual orbit.”

In regard to the theory, Borgondio was right; however, he had a 15-year delay. Flamsteed’s data raised much interest within the astronomical community soon after they were published. Early reactions in Britain were mostly positive; on the

[26]In Riccioli (1651, II, 452), this geometrical background remained implicit, but this attitude towards drawing “absurd” consequences in order to reject the Copernican system was already present (Baldini 1996; Baldini 1998, 744–747; Motta 2001).

[27]Dealing with a mathematical demonstration, Borgondio, in this case, did not use the astronomical notion of declination employed by Flamsteed; instead, he expressed himself in terms of inclination. Intuitively, two lines would be just as inclined toward one another as their extremes become closer, i.e., as the angular distance between them diminishes. Therefore, according to Borgondio, a star has greater inclination toward the terrestrial axis the higher it is on the horizon, i.e., as declination increases. In other words, Borgondio parametrized the inclination with respect to the terrestrial axis, whereas declination refers to the equatorial plane, normal to the axis.

continent, though, they were received with growing skepticism. By the turn of the century, the Italian-French astronomer Giovanni Domenico Cassini and the Dane Ole Rømer had expressed their reservations—the former publicly, the latter privately—and proposed that the maximum difference between the relative distances of the pole star from the Celestial North Pole (hence the difference in inclination toward the terrestrial axis, in Borgondio's terms) should occur between March or April and September or October, but not between June and December, as Flamsteed believed. Moreover, Cassini pointed out a mistake in Flamsteed's method, which he recognized in 1702 in a letter to Christopher Wren.

The idea that Borgondio was aware of these criticisms is improbable. First, Borgondio made only one explicit reference in the treatise—i.e., the citation of Flamsteed's letter in the *Opera* by Wallis, a book that was supposedly present in the library of the Roman College. He also mentioned some unnamed "French and English mathematicians" who found an annual variation in the position of the pole star over the years (Borgondio 1714, 6 and 10), but he did not provide other clues. However, there are only a few potential sources. Besides Flamsteed, he might have had in mind either the observations of the pole star by Philippe de La Hire and Jean Picard (a priest and former pupil of the Jesuits in La Flèche), as reported in the latter's *Voyage d'Uranibourg* (Paris: de l'Imprimerie Royale, 1680, 45–49), or those communicated by Cassini to the *Académie des Sciences* in 1693, of which he might have been aware.[28] In the second place, as mentioned, Rømer had expressed his criticism in a private letter to Flamsteed, unknown at that time, and Flamsteed's letter to Wren containing a frank admission of the mistake was published only after his death (1719). Third, and more importantly, neither Cassini nor Rømer doubted that the earth moved, but both challenge Flamsteed's interpretation of his data. They questioned whether what he had detected was the parallax of the pole star. Contrastingly, Borgondio was a committed anti-Copernican; he *did not* question Flamsteed's observations as such, which he took as genuine. He was only interested in showing that they did not match with the (geometrically correct) theory of the moving earth.

More generally, the problem was that the measurement of the annual parallax at that time was not conclusive at all (and so it remained, until Bessel measured the parallax of 61 Cygni in 1838). Borgondio's deductions from Copernicus' theory were correct; however, as Cassini and others suspected, Flamsteed did not observe the annual parallax. Therefore, against Borgondio's best hopes, his observations could not support the *hypothesis terrae motae*. In the end, Borgondio's mistake was that he was not as up-to-date as he thought. The title of his 1714 dissertation claimed that the motion of the earth was to be "assaulted [*impugnata*] by novel observations"; yet, these observations were already outdated.

[28]Cassini's observation, reported to the *Académie* in 1693, was published only later under the title "S'il est arrivé du changement dans la hauteur du Pôle, ou dans le cours du Soleil?", *Memoires de l'Académie royale des Sciences depuis 1666 jusqu'a 1699*, vol. X (Paris, 1730), 360–375. For a detailed historical account of the pole star observations in the context of the search for the stellar parallax, see Peters (1848, 8–14).

Within the Roman College, however, the urgent problem was to find counter-evidence against the Copernican theory. After all, since the times of Athanasius Kircher and his *Itinerarium exstaticum* (1656), Jesuits had searched for alternative explications of alleged parallactic phenomena (Siebert 2006, esp. 155–294). Therefore, it is likely that there were limited negative reactions, if any, to Borgondio's claims. His counterfactual-geometrical method was well accepted and, as we shall see, became a model in the early activity of his pupil, the young Ruggiero Boscovich, who succeeded Borgondio in 1740–41.[29]

1.4 Boscovich, the Anti-Copernican

It has been argued that late defenses of anti-Copernican positions of this kind, untenable in a world that was already Copernican, while Newtonianism was exuberantly entering the battleground, were, in fact, a malicious ploy designed to teach a disguised version of Copernicanism or, more generally, to convey the elements of officially forbidden approaches to the new generations of Jesuit scholars. These kinds of "rejections" would constitute a genre in their own right. In a resurgence of nicodemism, their authors were cryptocopernicans who, for many different reasons, were not ready to challenge a system of prohibitions and duties as sinister as that of the Roman Church. However, they had a long-term strategy that could bypass the threat: they could illustrate forbidden issues while publicly rejecting them.

Discussing this question in general would require the consideration of numerous sources, both manuscripted and printed, and would involve both historical and historiographical issues—which goes well beyond the aim of this book. For the present purposes, I restrict myself to taking Borgondio's dissertations (which would be relevant to Boscovich's early activity) at face value. Indeed, there is good reason

[29]As mentioned, Borgondio (1725) put forward another attempt to reject Copernicanism. In this case, the rejection occurred through an application of mechanics (typically, a *mathesis mixta*) to astronomy. If the earth moves around the sun, Borgondio reasoned, its motion can be treated in terms of the centrifugal forces involved. Therefore, he first presented the theorem of the centrifugal force (and illustrated it through a diagram); then, he applied it to the planetary case, investigated the ratio between the centrifugal force of the daily motion and that of the annual motion, found the ratio between gravity and the centrifugal force for daily motion, and finally "oppose[d] the system of the moving Earth through phenomena." Compared to Borgondio (1714), the 1725 dissertation is more general and provides a wider presentation of a Copernican-like system. Moreover, Huygens and, probably, Newton (whom he does not quote) are in the background. Huygens compared centrifugal force to gravity in his *Discours de la cause de la pesanteur*, published in 1690 (although it traces back to 1669), and showed that "if a body rotates along a circumference with the same velocity that it would acquire if it fell from a height of one quarter of the diameter, then centrifugal force and gravity would be equal" (see Bertoloni Meli, 1990, 26). Borgondio (1725, VII–VIII) reprised this result, and Huygens is quoted. He also recognized that, if the earth moves, by virtue of Kepler's third law, the centrifugal force of the planets is inversely proportional to the square of the distance from the sun (Borgondio 1725, IV—a result he might have known from Newton's *Principia*, Book I, Prop. IV, Coroll. 6 and Scholium).

to do so: nowhere did Borgondio provide general presentations of the Copernican system, nor did he offer accounts of particular aspects of heliocentrism. If he used Kepler's ellipses, he put them in a Tychonian context without presenting the heliocentric alternative. (In particular, no mention is made of Kepler's second law.) Finally, he did not appear up-to-date with the most advanced astronomical research, and, as we have seen, he quoted as "new observations" data that were already old.

With the anti-Copernican ban lasting until 1757, these aspects were also persistent when Boscovich took on the professorship of mathematics in 1740–41 (while Borgondio was appointed rector of the Roman College).[30] He had come back to Rome some years earlier, after a biennial stay in the small city of Fermo, located on a hilltop near the middle Adriatic Italian coast, where he taught *humanae litterae*. (This period was partly due to the poor condition of his health in 1734, which led the superiors to assign him to a Jesuit college in a more salubrious place.) At the Roman College again in 1736–37, he began to teach grammatics in the lower classes, then *humanae litterae* from 1737–38. It was during this occasion—at the inaugural lesson in November 1737—that he recited a didactic poem he had originally composed in 1735, while teaching and recovering in Fermo. A *carmen* of roughly 300 hexameters, it is, in fact, a short version of what would become the five-book-long (then six-book-long) poem *De Solis ac Lunae defectibus* ("On the eclipses of the Sun and the Moon").[31] The Fermo-version consisted of a preamble in prose and a *Pars prima* in verses; a *Pars secunda* was announced in the preamble as well, but there is evidence that it was never composed, so that the *carmen* in the projected form remained unfinished.[32]

---

[30]Ruggiero Boscovich announced this change in his life to his elder brother Natale (Bozo), in a letter dated February 27, 1740, apparently without enthusiasm but with some concern about the future: "Father Borgondio, Lecturer of mathematics, has been appointed as our Rector. He preserves his lectureship and people say that this [i.e., the post] is for me. In any event, this year I will have to replace him many times, and perhaps the next year too or always in the future or almost always" ("*Il P. Borgondio Lettore di Matemica è stato fatto nostro Rettore. Esso ritiene insieme la sua Lettura, e la gente dice, che sia per me. Qualunque cosa sia per essere, quest'anno mi tocchera supplire non poche volte, e forsi l'anno che viene, o sempre, o quasi sempre*" Boscovich 2012b, I, 46).

[31]Boscovich's (1735) early version of the poem of the eclipses is preserved as a manuscript in the Boscovich Archives at the Bancroft Library, Berkeley. It is entitled *De Solis, ac Lunae defectibus Carmen*, in which *carmen* means a medium-length composition.

[32]In the preface to *De Solis ac Lunae defectibus* (1760), Boscovich mentioned that the 1735 *carmen* was "roughly 300 verses long on the whole" (Boscovich 2012a, 32), which corresponds to the length of the manuscript. This circumstance is also consistent with a later note in Italian, with the autograph of Boscovich, on the page before the actual manuscript: "Poem of the Eclipses, as it was first composed in Fermo by Ruggiero Gius. Boscovich, in order to be recited in Rome, as he did on the inaugural lesson in November 1735 at the Roman College—which subsequently became a poem in 5 books, printed in London and Venice, and then in 6 books, reprinted in Paris, with a French translation" ("*Poema degli Eclissi, come fu composto da principio in Fermo/da Ruggiero Gius. Boscovich per recitarsi in Roma come fece nella/prefazione al cominciar a insegnare la prima in Collo Romano nel Nov: /del 1735 e che poi è divenuto di 5 libri stampato in Londra e in/Venezia,*

Boscovich introduced the poem with a clarification of the occasion and a declaration of the reasons that led him to the composition:

> The eclipse of the Moon on the 2nd of October, which appeared to me while, observing the sky, I was pondering over the subject of this inaugural lecture, troubled and hesitant, captured my attention; so I think I hardly could do something more appreciable to the most learned men than by dealing, in a heroic *carmen*, with "the Sun's many eclipses, the moon's many labors", which Virgil himself, the prince of poets, considers very apt to the 2nd book for the *Georgics*.[33]

The first part of the 1735 *carmen* addresses the causes and phenomenology of the eclipses, illustrates some related phenomena (e.g., the red color that appears in the terminal phase of a lunar eclipse), and refers to numerous theories. Thereafter, according to the preamble, the second part should present the motions of the sun and the moon in order to clarify "how the possibility of predicting their eclipses has to be explained." Finally—and surprisingly, in light of the Newtonian commitment transpiring from the published versions of the poem (1760/1778)—Boscovich announced, "I will also take the opportunity to explain by what manner the Copernican motion of the Earth has to be rejected."[34]

It is not entirely clear how Boscovich intended to manage the rejection of Copernicanism. A source might have been Riccioli (1651, II, 449–450), who exposed some arguments against the motion of the earth based on lunar eclipses and the evidence of the motion of the sun. In fact, it was Riccioli (1651, I, 286) who, quoting the same passage from the *Georgics*, suggested that the ambiguous term "*labores*" in Virgil's text should be interpreted as "*obscurationes*", i.e., occultations or eclipses. Riccioli even collected many quotations from classic authors, which

---

indi di 6, ristampato in Parigi nella traduzione francese"). For the history of the three versions of Boscovich's poem, see the Introduction to Boscovich (2012a, 18).

[33]"*Lunae defectus, quem mihi 2a Octobris nocte intueri licuit, iam diu de huiusce prolusionis argumento sollicitum, incertumque in eam adduxit mentem; ut nihil sapientissimis uiris me gratius facturum esse arbitrater, quam si, quos ipse poetarum Princeps Virgilius Georgicorum 2o carmini censet aptissimos, defectus Solis varios, Lunaque Labores, heroico carmine pertractandos susciperem.*" (Boscovich 1735, preamble, 1r, emphasis added). The (partial) lunar eclipse referred to by Boscovich took place on the night between October 1 and 2, 1735. He refers to *Georg*. II, 475–482: "*Me vero primum dulces ante omnia Musae, /Quarum sacra fero ingenti percussus amore, /Accipiant, caelique visa et sidera monstrent, /Defectus solis varios, lunaeque labores* . . ." ("But as for me—first may the Muses, sweet beyond compare, whose holy emblems, under the spell of mighty love, I bear, take me to themselves, and show me heaven's pathways, the stars, the sun's many eclipses, the moon's many labours." Virgil 1999, 171).

[34]"*In alteram vero partem Solis ac Lunae motus reieci, quorum Legibus cognitis, quo pacto praenunciari possint ipsi defectus explicandum, dataque occasione Copernicanus ille telluris motus refutandus esset.*" ("I have delayed the motions of the Sun and the Moon to the second part; their laws being known, I will also take the opportunity to explain by what manner the Copernican motion of the Earth has to be rejected." Boscovich 1735, preamble, 1r).

Boscovich used in his own hexameters, as proven by the marginalia to the 1735 *carmen*.[35]

The most important stimulus to the projected anti-Copernican attack arguably came from Borgondio's side. In the *carmen*, he is greeted as the one who introduced Boscovich to mathematics and astronomy. A didactic poet himself and (like Boscovich) a member of the Accademia degli Arcadi in Rome, he is mentioned after the arcadian name of "*Pastor Achemenides*" and celebrated for his contributions to didactic poetry and the mathematical investigation of nature.[36] Of course, this remained confined to the 1737 audience of the Roman College, since every reference to Borgondio was expunged from the editions published in 1760 and 1778. However, the grateful pupil would devote a *post mortem*-praise to his mentor in a long ecloga recited in 1753, and then issued 3 years later in a volume that collected poems in honor of deceased Arcadians (Borgondio died in 1741):

> The Shepherd Achemenides, drawing on wide tables,
> keeps the mind awake, o Lycida, and ravishes
> the spirit to him, and ask for grasping the quills.
> He, once upon a time, dragged me, seized with this,
> to the temples of holy mathematics, while he lied in the Goddess' presence,
> and he taught me to unfold the secrets of the concealed
> nature, and to observe the great ball of fire [i.e., the Sun]
> in the sky, and to watch through the clear nights.[37]

Boscovich had certainly known about Borgondio's attempted rejections of the *hypothesis terrae motae* and his revision of geostaticism through Kepler's ellipses.

---

[35]On Riccioli as a source of Boscovich's 1735 *carmen* and for a more detailed account of the differences between this early version and the published editions, see the Introduction to Boscovich (2012a).

[36]Boscovich revealed the identity of the Pastor Achemenide in a marginal note to the *carmen*: "Father Orazio Borgondio of the Society of Jesus, whose four poems stand out among the memories of the Arcadians: *De motu sanguinis, De incessu, De volatu, De natatu*" ("*Horatius Burgundius Soc: Iesu, cuius extant inter Arcadum monumenta poemata 4 de motu sanguinis, de incessu, de volatu, de natatu*"). His tribute in the original text of the 1735 *carmen* is as follows: "*Te primum mea vota petunt ... te in carmina pascunt/Pastor Achemenide Arcadie, ter maxima silvae/Gloria, dum sacras naturae inquirere leges .../Astrorum cursus idem, Coelique meatus/Scrutari vigil, et radio describere doctus/Ante alios. Tibi primus honos, tibi carminis huius/Debita laus omnis: Sub te [dedicisse] vigilare Magistro/Per sudum, puroque ignes scrutarier axe Matheseos artem/Capimus ethereos, sanctumque excolere. O quoties Lune spectare Labores/Vespere seu primo, media seu nocte per umbras/Perstitimus pariter! Tu tempora certa notabas / Dentatis inclusa rotis: Admoveras ipsum / Lente rubus vitrea insignis mihi comminus [docem].*" Boscovich 1735, 2r–2v. Note that, in the final version of the poem, all references to Borgondio have disappeared. On Borgondio and Boscovich in the context of the (Jesuit) "scientific poetry in Enlightenment Rome" see Haskell (2003, 179–244).

[37]"*Pastor Achemenides tabula pingendus in ampla/Sollicitum tenet, o Lycida, mentemque, animumque/Ad sese rapit, et calamos, dextramque reposcit./Ille olim sanctae correptum ad templa Mathesis/Me traxit, sistens Divae, atque arcana latentis/Pandere naturae, et magni scrutarier ignes/Aetheris, ac noctes docuit vigilare serenas*" (Boscovich 2012d, 28).

Possibly reminiscent of Borgondio's (1732) discussion of elliptical orbits within a Tychonic-like system is a dissertation that Boscovich devoted to the transit of Mercury across the sun on November 11, 1736.[38] This could have been a remarkable opportunity for possible considerations of the "chief world systems", such as in the case of Venus, as the observed phases of Mercury provided evidence of the motion of the planet around the sun. However, without debating the motion of the earth, Boscovich at first limited himself to presenting a theory of the motion of Mercury—in the case that it orbits the earth or the sun. Proposition VI of the treatise dealt with the case of a potential elliptic motion of Mercury having its focus in the sun. Here, Boscovich posed the problem of how to geometrically transform (an operation he calls *revocare*, i.e., to reduce or refer) "the geocentric motion of Mercury to a heliocentric motion," and the geometrical construction is performed (Boscovich 1737, IX–XIII). Only at this point did he seize the opportunity to note that "the resting state of the Earth [*Telluris quies*] does not exclude that the planets may follow elliptical orbits whose focus is the Sun, so that the sectors (inferred by the focus) are proportional to the times, and the squares of periodical times are proportional to the cubes of the semimajor axes" (Boscovich 1737, XIII, Cor. I to Prop. VI). With this, possible anti-geostatic objections seem to be neutralized—for the time being, at least.

A letter that he wrote on February 6, 1737, to his brother Bartolomeo (Baro) confirmed his geostatic/geocentric commitment in this period. In it, he presented a solution to an astronomical problem about which Baro might have asked his advice: "To find the curve to which all points that the Sun reaches during one year along one meridian belong."[39] Now, of course, tackling this problem implies the use of geocentric coordinates. However, if he had some Copernican inclinations, even if he could only have considered that the earth actually orbits the sun as a remote possibility, he probably would have made it clear that he was using some transformation from heliocentric coordinates to geocentric. But even if he did not see any need to do so, he would hardly have been content with simply stating that "the Solar orbit ... is elliptic according to the Moderns, with its focus in the Earth ..., but circular and yet eccentric according to the Ancients."[40]

[38]Titled *De Mercurii novissimo infra Solem Transitu* (Boscovich 1737), it was defended by three external students of the Roman College, but authored by Boscovich, as he reported in a letter to Giovan Stefano Conti on May 23, 1761 (Boscovich 2008, I, 45). Note that the dissertation was discussed in June 1737, only a few months before he recited the *carmen* on the eclipses.

[39]"*Trovare in che curva stanno tutti i punti ne' quali il Sole arriva in tutto un anno ad un Meridiano*" (Boscovich 2010a, 20).

[40]"*L'orbita solare ... è ellittica conforme i moderni col foco nella Terra ..., circolare conforme agl'antichi ma pure eccentrica alla Terra medesima*" (Boscovich 2010a, 20).

1.5 Exercising the Compatibilist Virtue

The "Moderns" had to be taken seriously, after all. It was perhaps in the early 1740s, shortly after he succeeded Borgondio as *professor matheseos*, that his awareness of the problems of the geostatic view developed considerably. He recognized two significant issues—stellar aberration and Newtonian gravitation—and discussed them on several occasions.

To begin with, what about stellar aberration? At some point between 1727 and 1728, James Bradley had discovered a periodic oscillation of 20 s of arc in the position of that star as he was attempting to confirm Robert Hooke's 1670s measurements of the annual parallax of γ Draconis. He also confirmed that other stars close to the zenith over his residence in Wanstead (northeast London) followed a similar pattern. Bradley concluded that the stars varied in terms of their positions, though not in the manner that Hooke explicated through a parallactic displacement. Rather, the variation was due to an "aberration of light." Assuming a finite speed for the light, a beam of light coming from a star must take a finite amount of time to travel through the telescope—a brief time interval during which, however, the tube has moved slightly because of the orbital motion of the earth. In other words, the phenomenon that Hooke observed was not the stellar parallax (the annual displacement of the star caused by the displacement of the observing location); it was, in fact, the aberration of light, but required a moving earth just as much.[41]

Though he was convinced that light propagates with finite speed and that Bradley's aberration actually takes place, Boscovich was not willing to make concessions about Earth's motion. His attack on the problem is contained in a 1742 dissertation entirely devoted to the "stellar annual aberrations." It is worthwhile noting that Boscovich's usage of this phrase is slightly different from the common meaning. Whereas 'aberration' usually refers to the phenomenon identified by Bradley while 'parallax change' is conceived as an essentially distinct event, he tended to employ *aberratio* almost in an etymological sense, meaning any deviation from a certain value assumed to be true or standard (see also Boscovich 1742c, §§ 7, 37). So, he recognized that certain annual variations in the stellar position, which he globally labelled *aberrationes*, "are known to everyone" (Boscovich 1742a, § 1): He referred to the search for stellar parallaxis in general, the discovery of a parallactic displacement by Hooke and Flamsteed, and its explication in terms of light aberration by Bradley. However, Boscovich argued that the motion of the earth does not necessarily follow from Bradley's aberration and that a plurality of candidates can explain it just as well. As he stated, "for the sake of astronomical exercise . . . we select one of these, which is especially in accordance with the other constitution of the system of the world, where the Earth is stationary" (Boscovich 1742a, § 2).

He started allowing, with Tycho, for three kinds of motion:

[41]On Bradley's aberration and its discovery, see Hoskin (1982, 32–35), Heterington (1988, 19–21), Taton and Wilson (1989, 156–157), Hoskin (1997, 206, 212–215), Gualandi and Bònoli (2004, 2009), Fisher (2010).

> (1) The common motion, that is, the motion by which every day celestial bodies are brought from east to west around the equatorial axis. Associated with it [*cui affinis*] is the very slow and simultaneous motion of all celestial bodies around the ecliptic axis toward east ... (2) The proper motion of the secondary planets [i.e., satellites] around the primary and that of the primary planets and the comets ... around the sun, in the same orbits and with the same velocities than in the hypothesis of the moving earth. (3) The annual motion of the sun around the earth in an orbit that is equal to that described by the earth around the sun, but with mutually inverted positions. All remaining orbits of planets and comets except the moon, moving together with the sun, accompany this motion, so that the line joining any point of them with the center of the sun preserves the size and symmetry [*magnitudinem, et parallelismum servet*]. (*ibid*., § 3)

As already noted, Tycho's and Copernicus' systems could be viewed as geometrically equivalent. However, both the parallactic displacement and Bradley's aberration of light represented discrepancies between them, strongly threatening their symmetry. In this sense, stellar parallax and the aberration of light can be viewed as the ultimate crucial observations capable of deciding between geocentrism (*hypothesis terrae quiescentis*) and heliocentrism (*hypothesis terrae motae*). Nevertheless, perhaps a last way out was left to the stubborn geocentrist. Indeed, a problem with systems like Tycho's was that they conceive of the sphere of fixed stars as being independent from the motion of the sun. At this point, Boscovich introduced a remarkable auxiliary hypothesis in order to preserve the equivalence between Tycho-styled systems and the cosmology of the moving earth:

> We suppose [*ponimus*] that not only the orbits of the planets and those of the comets, but also everything concerning the fixed stars accompany the annual motion of the sun, so that the whole sphere reaching from the sun, including all bodies except the moon, would move around the earth, with annual motion capable of preserving symmetry, as it would be a sole system. (*ibid*.)

The "astronomical exercise" discusses two series of three cases each. The first group, with the earth in annual motion around the sun, is composed as follows: "(1) Light propagates instantaneously and parallax is appreciable. (2) Light propagates as time passes [*Propagatio sit successive*] and no parallax is perceptible because of the immense distance of the fixed stars. (3) Light propagates as time passes and parallax is appreciable." The second group, with a stationary earth around which the sun is moving and stars move accordingly, includes: "(4) Light propagates instantaneously. (5) Starlight reaches the earth in 3 months. (6) Starlight takes a shorter interval of time to reach the earth."[42] Boscovich contends that, in all of these cases — therefore in the *hypothesis terrae motae*, as well as in the *hypothesis terrae quiescentis*—"any fixed star whatsoever apparently describes an ellipse (approximately or precisely)" and the same phenomena are observed "in case 1 and 4; in cases 2 and five; in cases 3 and 6." (*ibid*., 1742a, § 4).

[42]Boscovich did not explain why he assumed that light should take up to 3 months to reach the earth from any star whatsoever. To be sure, he knew that this was not a plausible hypothesis at that time. Probably, the 3 month-pace is only assumed so as to simplify the calculation of case 5 (I will come back to this).

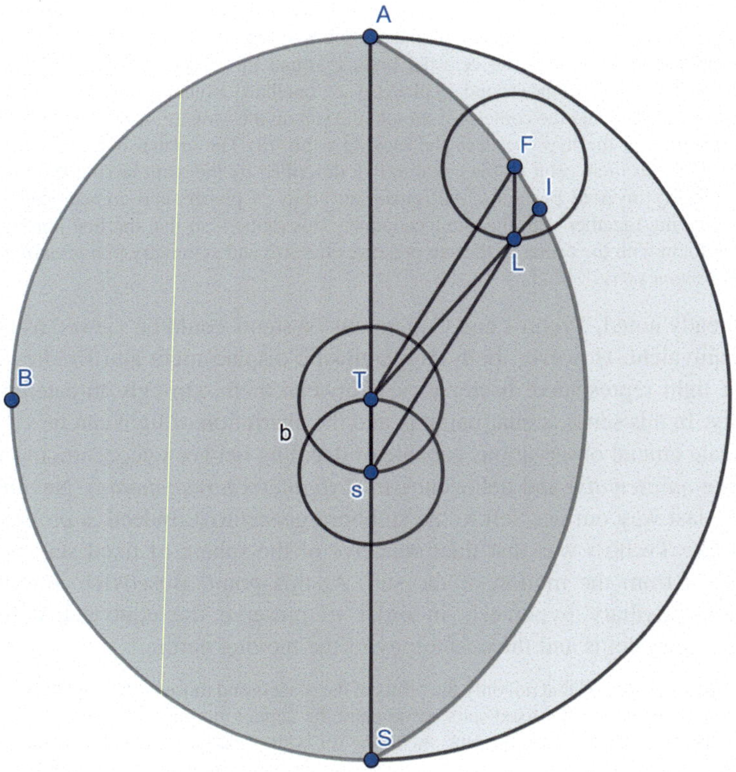

Fig. 1.3 Boscovich's basic diagram for comparing aberrations in the *hypothesis terrae motae* vs. the *hypothesis terrae quiescentis*. In particular, this applies to the case of parallactic displacement, *s* being the sun and *S* its projection on the ecliptic ABS. *I* is the apparent position of a fixed star *L* on the celestial sphere, for an observer placed on the earth *T*. Because of the parallax, *I* is seen as moving around the point *F* and *FI* is the "absolute aberration by parallax" (Boscovich 1742a, § 13: let us remember that he does not employ *aberration* in the usual sense, but with a wider scope, meaning any deviation from a standard value). Note that the true direction *TL*, which is the joining line between the observer and the star, coincides with the apparent direction *TI*, which is the direction along which the object is actually observed on the celestial sphere. (The figure is a modern adaptation from Fig. 2 of Boscovich 1742a; I did my best to make it clear, while preserving the details of Boscovich's original diagrams, that the arcs represented here define sections of a sphere)

Boscovich's demonstration (see *ibid*., §§ 12–24) is somewhat long and complicated, so the next discussion will be limited to its basic elements. We want to solve the problem of determining the apparent orbit of a fixed star *L* with respect to the earth. Let us consider Fig. 1.3, where [stet] *s* is the sun, *T* the earth, *sL* the joining line between the sun and the fixed star, and *bs* and *bT*, respectively, the opposite and equal orbits of the sun around the earth and the earth around the sun. Imagine constructing the celestial sphere (upon which objects in the sky can be conceived as projected): it is an abstract sphere centered on the earth, having an arbitrarily large radius. Suppose that *T* is moving around the (eccentric) *s* in the approximately circular orbit bT; the sphere moves as well. Now, let us take the radius TF so that it is parallel and equal to sL. Finally, the ecliptic plane Tbs intersects the sphere in its great circle ABS, which is the ecliptic.

TF and sL being equal and parallel, the segment FL will also be equal and parallel to the distance Ts of the earth from the sun. Let us imagine that the sun and the star are stationary, one in respect to the other: by virtue of the motion of the sphere along with that of the earth, sL will describe a circle around F—an "orbit" that is parallel and equal to the apparent motion of s around T. Now, let us consider how an observer in T will perceive L projected onto the surface of the sphere. Imagine joining a terrestrial observer T with L and extending the line until it intersects the sphere in I, where it will appear; the motion of TL will produce a conic surface with axis TF. Since the distance of the surface of the sphere is assumed to be very large, to a terrestrial observer, this will appear like a plane surface that cuts the cone generated by TL at a certain angle, resulting in an ellipse. In other words, the fixed star L will be perceived by an observer placed in T as moving along an elliptical orbit, with position L projected as I on the celestial sphere.

In the first case (earth moving, instantaneous propagation of light, appreciable parallax), "the fixed star L observed from earth T will appear in I, in accordance with the direction TLI" (*ibid.*, § 12). There will be no Bradley aberration, but only an "absolute aberration descending from parallax" [*aberratio absoluta orta ex parallaxi*] (*ibid.*, § 13). As can be easily seen, it corresponds to the fourth case (earth stationary, instantaneous propagation of light). However, whereas, in the explication of the figure above, we first supposed that an external observer (say, looking from God's privileged eye) would see an earth in motion around a stationary system composed by the sun and the stars, as if rigid rods connect them to one another, now God's eye is supposed to observe the same sun-stars system as if it were in motion around a stationary earth: "In the fourth case, the earth, as well as the sphere, being stationary, whereas the sun s is moving on the orbit sb along with sL, and all other things remaining equal, FL will describe the same orbit equal to bs, the line TIL will describe the same conic surface, and thus point I will describe the same ellipse on the surface of the sphere" (*ibid.*, § 14).

Before considering case 2, let us remark upon the following: if, in Fig. 1.3, the relative distance between the terrestrial, approximately circular orbit bT and the fixed star L are thought to be so enormous that earth's annual orbit is "like a point, if compared with such a distance" (*ibid.*, § 15), any stellar parallax will disappear and, *if light propagates instantaneously*, it will always be seen in the position F on the sphere. In other words, because of the distances involved, the elliptical path of I will also be seen "like a point", and the displacement of I from F, or *aberratio ex parallaxi*, will be vanishingly small, and thus imperceptible.

However, if light propagates as time passes, things will be different, as illustrated in Fig. 1.4. So, let us take into consideration case 2, in which light proceeds with finite speed but no parallax occurs. Suppose, as it is, that the ratio between the speed c of light and the velocity v of the Earth along its orbit is appreciable. Then, "let us take, on the line AD tangent to the orbit bT in T, the segment TD, so that $TD : TF = v : c$,

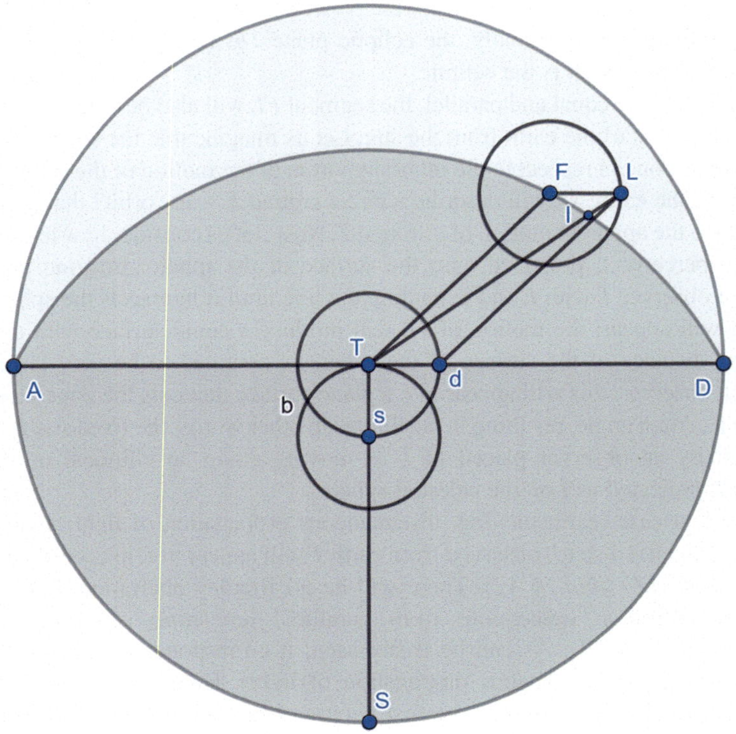

Fig. 1.4 Boscovich's diagram for the case of Bradley's aberration of light. Let *TF* be the true direction. *L* will be seen in *I* as the combined effect of the relative motion of the earth and the star and the speed of light with respect to the velocity of the earth on its orbit. Note that the direction of the aberration is different from the direction in which the parallax occurred in Fig. 1.3. (The figure is a modern adaptation from Fig. 3 of Boscovich 1742a; as in the previous case, the arcs represented here define sections of the celestial sphere)

and let us draw *dL* so that the parallelogram *TFLd* is complete" (*ibid.*). According to the rules of trigonometry, we have the following relations:

$$\frac{\sin d\mathrm{TL}}{\sin dLT} = \frac{\sin dTL}{\sin LTF} = \frac{dL}{TD} = \frac{TF}{TD} = \frac{c}{v}$$

Let *TF* be the true direction of the star (that is, the joining line between a fixed star and the observer at any instant of time) and *ATD* the horizontal axis of a terrestrial observer's gaze; then, *TL* is the apparent direction, as the combined effect of the Earth's motion and the speed of light. Moreover, if the ratio $\frac{c}{v}$ is assumed to be constant (which is false, but reasonable, because *T* moves on an ellipse that is approximately a circle and, according to Kepler's second law, in the case of a circular orbit, a planet would move with constant velocity), then *TD* and *FL* will also be constant and point *L* will describe an approximately circular path around *F*, *TL* will describe a conic surface in the sphere, and *I* will result following an elliptic orbit—as in the case of parallactic displacement. However, the direction in which the aberration occurs is different. In the case of parallax (Fig. 1.3), it was directed toward the projection *S* of the sun *s* on the ecliptic; now, with the speed of light becoming relevant, it is the earth's (velocity of) motion relative to (the velocity of) light that matters. So, in Fig. 1.4, the direction of aberration ("*punctum dirigens aberrationem*") is given by point *D*,[43] the eastern point on the ecliptic.

Now, let us consider the *hypothesis terrae quiescientis*, thus assuming that the earth is stationary and the sun revolves around it from east to west in the annual orbit *sbd*, where *ds* is a quarter of the orbit. Therefore, the sun will take 3 months to recover, starting from *d*, position *s* with respect to the earth *T*—or, the great circle *ADS* being the ecliptic, the sun will take 3 months to appear in *S* starting from *D*. Now suppose, as in case 5, that the light from star *L* takes 3 months to arrive at the terrestrial observer's eye in *T* (this is an arbitrary hypothesis for the sake of simplicity, 3 months being a quarter of 1 year, i.e., the time that the sun takes to cover a distance like *ds*). With the sun in *s*, the star will appear in *I* and "will always be seen, like in case 4, as travelling an ellipse perpetually passing through *I*" (*ibid.*, § 17) However, it will obviously be seen in the position that it had when light was emitted, hence 3 months earlier—that is, when the sun was seen in *D*, which is the direction of aberration, as in case 2 of the *hypothesis terrae motae*.

This geostatic description of aberration can be generalized considering shorter times for light propagation (or, allegedly, longer times), and can also be combined with the parallactic displacement, as Boscovich shows in the subsequent paragraphs, contrasting case 3 with 6. For any heliostatic case, he suggested, its geostatic equivalent can be provided. Is it the sun or the earth that is really moving? For an observer placed within the system represented by Boscovich's diagram, there is no privileged frame of reference, all observed phenomena being equal. In this sense, as he concluded, "this whole investigation depends on hypotheses and, whenever one does not deal with appearances but with their causes, nothing can be demonstrated" (Boscovich 1742a, 19). For Boscovich, Bradley's aberration and parallactic displacement were but cases of theoretical underdetermination that can easily be made compatible with a stationary earth.

---

[43]Boscovich's language can be somewhat misleading. Let us remember that, in both the case of the parallax shift and that of the aberration of light, the observed positions of the star describe an annual ellipse with the true position at its center. Moreover, in both cases, the ellipses are traversed according to the same direction, but with one quadrant off-kilter. Note that Boscovich formally defines the *punctum dirigens aberrationem*, both in the case of the "absolute aberration by parallaxis" and in the proper case of aberration of light, as "the point of intersection, on the ecliptic ..., with an arc of great circle drawn from the intersection of the true direction with the spheric surface along the intersection with the apparent direction". So, in Fig. 1.3, it is defined by the arc *FIS* (*TF* and *TI* being the true and the apparent directions, respectively), whereas, in Fig. 1.4, it is defined by *FID*.

1.6 Glimmers of Newtonianism

In the following years, he learned that he was underrating their potential, particularly in the case in which they were coupled with other weapons. Apparently, he prosecuted on his defensive line. In the same year in which he published his Tycho-styled solution of these anomalies—the aberration of light and the displacement of stellar parallax—Boscovich also provided an overview of astronomy in a work titled *Disquisitio in universam astronomiam* (Boscovich 1742c). Here, he emphasized a noticeable agreement between observations and "the Keplerian hypothesis" *of the elliptical orbits, which, he claimed, surpassed that of any other theory. He offered the following explanation:*

> *Kepler's laws* can be adjusted to a stationary Earth as well, so that the remaining primary planets would move around the Sun on elliptical paths, the orbit of the Earth would be substituted by another orbit of equal and contrary order in which the Sun would lead around the orbits of the planets in the same way as Saturn and Jupiter lead around their satellites . . . The hypothesis of the moving Earth is somewhat easier and more elegant; still, even this simplicity—leaving aside the Holy Scriptures—does not suffice to settle the dispute, and the idea underlying the stationary Earth, which we defend, has a certain analogy too and is not inelegant. (Boscovich 1742c, 15–16)

By 1742, he must have already realized that this kind of naïve compatibilism was a strategy without a future and, as such, did not go beyond the proclamation. This was probably a consequence of his increasing understanding of Newton's *Principia*. It is not clear when, exactly, he read the treatise, but his discussions of Newton's theory seem to occur in connection with his geodesic interests between the late 1730s and the early 1740s.[44] In this period, he authored a couple of dissertations about the shape of the earth and a treatise about gravitational inequalities at different locations on the terrestrial surface. In each of these works, Boscovich repeatedly stated that empirical evidence from measurements carried out on the earth is not enough to support the *hypothesis Terrae motae*. Moreover, such measurements could support more than one theory of gravity.[45] It is in this context that he first referred to Newton's concept of gravity as decreasing with the inverse square of the distance:

> Finally . . ., the mutual attraction according to the inverse square of the distance is a hypothesis. It is certainly a very ingenuous one, and derived with astonishing felicity from

[44]This happened, therefore, shortly before the 1742 astronomical exercise commented upon in the previous section was issued. Note that, here, he claimed that, even in his hypothesis of a Tychonic-styled system, "the physical cause of the motions . . . can be found once the Newtonian attraction is admitted [*admissa tantum Newtoniana attractione*]". But he immediately specified that "this matter is not appropriate to the present discussion [*sed ea non sunt hujus loci*]" (Boscovich 1742a, § 2).

[45]For example, Boscovich (1739b, XXXIII) claimed that observations do not suffice to prove Newton's assumption, whereas other hypotheses produced slight anomalies "even if we assume that the Earth is moving." The three dissertations are titled, respectively, *De veterum argumentis pro telluris sphaericitate* (Boscovich 1739a), *De figura telluris* (Boscovich 1739b), and *De inaequalitate gravitatis in diversis terrae locis* (Boscovich 1741a).

Kepler's rules, in which it dealt with the planets, and the Earth is included in their number; nonetheless, it is highly arbitrary and mechanically inexplicable when applied to the tiniest particles of matter that attract each other according to the same law. . . . We hold that nothing ingenuous [see above] has been conceived so far, nothing more consistent with geometry and observation, than the entire bulk and structure of Newton's system. However, we also hold that, as far as it assumes the motion of the Earth, no geometer will deny that this is a pure hypothesis.[46]

In spite of all of the possible specifications regarding what is, in fact, a hypothesis in Newton's vocabulary (and, to be sure, Newton would deny that the inverse-square law is such a hypothesis), this judgment about the convenient-but-conjectural nature of Newtonian gravity was repeated in 1742 with an emphasis on the astronomical aspects. Let us see, first of all, why it is convenient: if we "assume that all particles of matter attract each other according to the inverse square of the distance, and also that the planets were initially thrown with due velocities, then not only can Kepler's laws be well deduced with geometrical rigor, but almost the entire astronomy follows almost spontaneously. Indeed, the Lunar anomalies, which for a long time have escaped the control of any law, are described with the best approximation, and the motion of the apsides, the precession of the equinoxes, as well as the orbits of the comets are happily deduced and consistent with the observations. It is also most convenient that, from the mutual action of the planets, it can be explained why the observations do not altogether agree, why the planetary tables, when a longer timespan is considered, gradually differ from the sky, so that they cause remarkable errors, if they are not restored through new observations" (Boscovich 1742c, 21, § 54). And a further advantage of Newton's conception is that one is led to dispense with Cartesian vortex theory---something that Boscovich "absolutely [did] not like" (ibid., 16, §§ 39-40; see also Boscovich 1741a, § 29). However, a threefold weakness of Newton's theory was identified: "(1) It is a hypothesis that cannot be immediately deduced from pure phenomena. (2) It cannot be demonstrated as completely agreeing with phenomena. (3) It has its own difficulties, which are not negligible." (Boscovich 1742c, 21, § 54) The final paragraphs of this treatise (ibid., §§ 55–59) are devoted to substantiating such deficiencies. Perhaps this was only instrumental in dissimulating the potential of Newton's theory in solving actual problems. However, Boscovich (ibid., 22, § 58) referred to two real flaws, which he described as "*duae gravissimae difficultates*": the first flaw concerns the

[46]"*Quid si tandem accedat, ipsam attractionem in ratione reciproca duplicata distantiarum hypothesim esse, ingeniosissimam sane, et mira felicitate erutam ex Keplerianis regulis, ubi agitur de Planetis, et inter Planetas numeratur Tellus; maxime tamen arbitrariam, et mechanice inexplicabilem, ubi agitur de minimis materiae particulis se mutuo attrahentibus in eadem lege? . . . Putamus quidem nihil hactenus excogitatum esse ingeniosius, nihil cum Geometria, et observationibus consentaneum magis, quam sit universa Newtoniani systematis moles, atque structura. Eam tamen etiam posito telluris motu, puram esse hypothesim, nemo, ut arbitramur, Geometra abnuet*" (Boscovich 1741a, §§ 27, 29).

cosmological paradox and the second, possibly reminiscent of the "mechanical inexplicability" mentioned in the above-quoted passage from Boscovich (1741a), is the causal mechanism of gravitation.[47]

Most importantly, here, Boscovich seemed aware that, in spite of any proclamation about compatibility, in principle, with Kepler's laws, a stationary earth *cannot be adjusted* to Newton's system—at least not in the terms of the geometrical compatibilism that he had pursued. In his explanation of why he maintained that Newton's concept of gravity "is a hypothesis that cannot be immediately deduced from pure phenomena," Boscovich asserted the following:

> From the uniform description of the areas in an ellipse around its focus, Newton deduces an accelerative force of the secondary planets [i.e., the satellites] towards the primary planets, and of the primary planets towards the Sun, which is proportional to the inverse square of the distances if the distances of the same planet are compared. Moreover, he draws out this same law from [Kepler's] third law, if distances pertaining to different planets are compared. *However, first of all, if the Earth is stationary, the daily motion [of the planets] would greatly perturb the description of uniform areas.* Therefore, if the motion of the Earth is not immediately deduced from the observations—and certainly it is not—the main element of [Newton's] theory will not be deduced as well.[48]

What he was pointing to with the remark emphasized above is not entirely clear, but some conjectures seem plausible.

As a skilled reader of *The Principia*, he could have hardly overlooked "Phenomenon 4", where Newton, commenting on Kepler's third law, remarked that "the

[47]"Finally, there have always been, and there will always be, two extremely severe difficulties .... As to the first, the fixed stars would rush one into another because of their mutual action, and would perpetually accelerate to form a single mass—unless God is employed as a machine to keep them back. As to the second flaw, an action should be ascribed to the particles of matter, so that they act at the farthest and, of course, most immense distances without any medium by which they act, and for whose action no mechanical cause can be contrived—unless we appeal either to the occult nature of things or to God's free law, as itself being superior to the individual laws." ["*Demum fuerunt semper, eruntque illae duae gravissimae difficultates, quarum posterios maxime sane multos absterruit semper, et adhuc ab eo complectendo systemate absterret. Prima, quod Fixae vi mutuae actionis in se invicem irruerent, acad unam massam efformandam perpetuo properarent, nisi Deus ut machina ad eas cohibendas adhibeatur: Secunda, quod action tribui debeat particulis materiae, quae agant in maximis ac plane immensis distantiis sine ullo medio, cujus ope agant, et cujus actionis nulla causa mec[h]anica excogitari possit, nisi vel ad occultas rerum naturas, vel ad liberam Dei legem provocemus singular per se praestantem.*" Boscovich 1742c, 22–23, § 58.] This possibly reveals that Boscovich directly or indirectly knew of Bentley's exchange with Newton, where both questions are raised (see, for example, Newton, 2004, 94–105). As we shall see, Boscovich's curve would entail an attempt to solve the cosmological paradox. See Boscovich (1763, §§ 398–400); also see Guicciardini (1996, 276–277).

[48]"*Deducit quidem Newtonus ex aequabili descriptione arearum in Ellipsi circa ejus focum, vim acceleratricem Secundariorum i[n] Primarios, et Primariorum cum Secundariis in Solem, quae sit in ratione reciproca duplicata diftantiarum, si ejusdem Planetae distantiae inter se conferantur, et eandem legem, si diversi Planetae conferantur inter se, eruit ex tertia lege.* Sed in primis si Terra stet, arearum aequabilem descriptionem turbat plurimum diurnus motus. *Quare si immediate ex observationibus non deducatur Terrae motus, qui certe non deducitur, non deducetur praecipuum elementum ejus theoriae*" (Boscovich 1742c, 21, § 55, emphasis added).

periodic times are the same, and the dimensions of the orbits are the same, whether the sun revolves about the earth, or the earth about the sun" (see Newton 1687, 800). Of course, Ptolemy's system was out of the question (because of, among other reasons, the phases of Venus); after all, Newton (1687, 799) explicitly excluded it in the preceding "Phenomenon 3. The orbits of the five primary planets—Mercury, Venus, Mars, Jupiter, and Saturn—encircle the Sun." But this still left room for a kinematic equivalence of a Tychonian-based cosmology with Newton's system, which was indeed—and perhaps Boscovich intended to point to this feature—only a kinematic equivalence. In other words, Boscovich (1742c) seemed to have realized that, once masses and gravitation are taken into account, no equivalence between Newton's system and Tycho's system is possible.

Newton's gravitation is not Bradley's aberration: they have a different epistemic status. Aberration is but *one* theory that explains an empirical fact (an annual displacement of the stars); to be sure, Boscovich (1742a) accepted the empirical fact, but not the theory. According to Bradley, the causes of the stellar displacement are the orbital motion of the earth *and* the finite speed of light. Or, from the hypothesis of the orbital motion of the earth *and* that of the finite speed of light, one can derive his theory of aberration: *If* Earth's motion and finite speed of light, *then* aberration for explaining stellar displacement. Boscovich retorted that he could easily find other explanations for the same empirical fact, starting with the hypothesis of a motionless, Tychonic earth. Gravitation, however, is a "hypothesis" (for Boscovich) in its own right—a more basic assumption than aberration. Newton noted that this is simply the way bodies behave. There is no *if. . .then. . .* as in the case of aberration; there is, instead, an asserted state of affairs.

However (or, rather, exactly for this reason), Newton's universal gravitation represents a much harsher epistemic constraint. Once it is accepted—and, starting with Boscovich (1742c), he was inclined to do this—*any* two masses in the Universe attract each other. In other words, let either mass accelerate the other mass, and the acceleration should decrease proportional to the square of the distance. Now, let the center of the Universe be occupied by a fixed earth (or, more precisely, the earth should revolve around it in an ellipse, however small); the sun, however, moves in a large Keplerian ellipse around the earth (no matter what the fate of the other planets would be). If Newton's law is observed, Kepler's laws are observed as well; from a kinematical point of view, the above-quoted "Phenomenon 4" from the *Principia* applies: The periodic times are the same, and the dimensions of the orbits are the same as in Newton's system, taking for granted that it is the sun that orbits the earth. To allow for this, however, the earth's mass should be many thousands of times that of the sun. Under such conditions, when the other planets are included in the system, they must revolve around a featherweight sun in very small ellipses. (Earth's gravity would otherwise capture them.) Probably for this reason, Boscovich noticed that, under the assumption of a stationary earth *and* Newton's law, some discrepancies with the observed daily motions of the planets would occur, and no equivalence between the *hypothesis terrae quiescentis* and its nemesis—the moving earth—can be provided.

In the early 1740s, the Newtonian commitment that would characterize much of Boscovich's later career, ending up as a "sort of apotheosis" of Newton in the fifth book of *De solis ac lunae defectibus* (1760), was smoldering under the ashes of the compatibilistic strategy that he had previously endorsed.[49] Newton's theory of gravitation contributed to changing his mind. His interest in it first developed as a consequence of his approach to issues specifically related to geodesy, but he soon realized that, if Newton was right about gravity, this could have a tremendous impact on cosmology: The kinematic equivalence by virtue of geometry, on which his own (and other confrères') compatibilism was based, appeared out of the question. He had to change his strategy if he wanted to comply with the Jesuit doctrine of a stationary earth.

1.7 Compatibilism Anew: The "Sidereal Space"

From that point on, the name Newton began to appear more frequently in Boscovich's papers. As we shall see, the reference to Newton would play an important role in the mechanical dissertations of the early 1740s, as well as in the 1745 *De viribus vivis*, in which Boscovich first formulated his own law of attraction and repulsion. And, of course, if this law holds, Earth cannot be physically immovable, since attraction is mutual.

The process of acquaintance with the Newtonian system, however, did not proceed as smoothly as it might retrospectively appear *prima facie*. In a 1746 dissertation on comets, he declared his sympathy for Newton from the very beginning in statements such as the following: "Comets are perennial stars that follow the general laws of motion together with the planets (although they differ in many things from them), and this occurs according to the theory of Newton" (Boscovich 1746, §2). Nonetheless, he continued, though such a theory assumes that Earth is moving, we do the following:

> We, revering the testimony of the Sacred Scripture and complying with the decree of the Holy Roman Inquisition, hold [*statuimus*] it as immovable. We consider its motion only as regards the appearance [*in speciem tantum*] for the sake of an easier description, demonstrating that absolutely the same phenomena occur either in the case that the Earth moves

[49]This attitude is perhaps instantiated at its best in the cautionary concluding remark to Boscovich (1742a), after he expounded his doubts about Newton's system: "From all this it is evident to what extent we should be careful . . . in considering as a very seldom case, and not without difficulty, that in physical and physical-mathematical arguments only the physical demonstration matters." In fact, mathematics also matters—in particular, calculation joined with empirical evidence in order to confirm or reject any particular theory. However, the difficulties that Boscovich had in mind are unclear when he ambiguously stated "from all this." Does he mean the displayed problems with Newton's theory? Or, is he instead pointing to the problems with the *hypothesis terrae quiescientis*, "which we defend" even if it is less simple and elegant (Boscovich 1742a, § 38). Is he intimating that the theory of a stationary earth displays anomalies when dealing with the daily motion of the planets (§ 39) and is not as efficient as Newton's theory in solving other tasks (§ 54)?

> around the Sun or in the case that the cometary orbits, together with the Sun, revolve around the motionless Earth. (Boscovich 1746, §2)

This only apparently signals the usual "weak compatibilist scenario" conforming to the old Bellarminian instrumentalist dictate. In fact, Boscovich's words take a remarkably semantic turn. The earth's motion is considered *in speciem tantum*, "only as regards the external aspect, the appearance." However, note that the term *species* means something different than the rather technical concept of *apparentiae* [appearances], which usually occurs in the plural form, in astronomy. When Clavius (1611b, 301), for example, stated that "from eccentric orbs and epicycles not only are the appearances [*apparentiae*] already known defended, but the future ones are also predicted," he implied that the Ptolemaic model satisfactorily saves and predicts the observed processes, i.e., the apparent motions of the sun, the moon, and the planets. In his letter to Foscarini (April 12, 1615), Bellarmino added that, in order to better describe such appearances, one can feign that the sun is motionless and that the earth revolves around it. However, as he also commented, "it is not the same to demonstrate that, by supposing the sun to be at the center and the earth in heaven, one can save the appearances [*si salvino le apparenze*], and to demonstrate that in truth the sun is at the center and the earth in heaven; for I believe the first demonstration may be available, but I have very great doubts about the second."[50]

According to Bellarmino, the sun is *physically* moving and the earth is *physically* at rest. To dispel all doubts, he rebutted a potential objection:

> Suppose you say that Solomon [quoting from *Eccl.* 1,5: "The sun rises, and the Sun goes down, and hastens to the place where it rises"] speaks in accordance with the appearance [*parlò secondo l'apparenza*], since it seems to us that the Sun moves (while the earth does so), just as, to someone who moves away from the seashore on a ship, it looks like the shore is moving. I shall answer that, when someone moves away from the shore, although it appears to him that the shore is moving away from him, nevertheless he knows that this is an error and corrects it, seeing clearly that the ship moves and not the shore; but in regard to the Sun and the Earth, no scientist [*savio*] has any need to correct the error, since he clearly experiences that the Earth stands still and that the eye is not in error when it judges that the Sun moves, as it also is not in error when it judges that the Moon and the stars move. (Bellarmino to Foscarini, 12 April 1615; quoted after Finocchiaro 1989, 68–69.)

Note that, here, the singular Italian noun "*apparenza*" (appearance) does not have the same semantic value as the plural form Bellarmino employed in the phrase "*si salvino le apparenze*" (one can save the appearances). In that case, the term is rather technical and refers to the observed phenomena, more or less in the sense instantiated by Clavius in the quoted passage above. However, the expression "*secondo*

[50]Bellarmino's letter to Foscarini has been pubished by Favaro in Galilei's collected works (*Le Opere di Galileo Galilei*, Edizione Nazionale, ed. by A. Favaro, Firenze: Barbera 1938, vol. 12: 171–172). This and the following passage are mainly quoted after the version of Finocchiaro (1989, 68–69), possibly with slight changes in the translation.

l'apparenza" refers to what someone sees though what is seen is not physically real.[51]

Of course, Boscovich's expression "*in speciem tantum*" might reflect the philosophical vocabulary of the Order. However, its application to the astronomical context is reminiscent of Bellarmino's usage of the phrase "*secondo l'apparenza.*" In other words, by considering the motion of the earth *in speciem tantum*, Boscovich played the role of Bellarmino's impossible *savio*. That the earth is moving *in speciem tantum* means that—paraphrasing Bellarmino's letter—an astronomically aware observer, knowing Newton's law and some other explications (e.g., Bradley's aberration), clearly experiences that the earth physically moves. However, the *savio* knows that, in truth, the earth stands still and all of the other things revolve around it. In fact, Boscovich (1746, §11) clarified that the motion of the earth is "relative to a space that we conceive of as translated together with the Sun. When we state that the earth has moved from *T* to *t*, we mean that point *t* of a certain space, which we conceive of as being in motion together with the sun, has moved to *T*."[52]

This conception—which is reminiscent of the "symmetric system" quoted above, from Boscovich (1742a, § 3)—was exposed in full detail a year later, in a treatise devoted to the sea-tides (*Dissertatio de maris aestu*: sea Boscovich 1747). Having presented Newton's theory of the tides, Boscovich (1747, §§ 68–69) quoted the 1746 dissertation on the comets and attempted to show "in which manner this theory and the whole Newtonian physics can be reconciled with the resting Earth." Let us consider a "sidereal space" (*Spatium sydereum*) that contains all the fixed stars, the planets, and the comets—briefly, the entire observable Universe. Let every body in it be endowed with inertia relative to that space, as well as with the forces of the usual Newtonian mechanics, including gravitation. Now, imagine holding the earth at rest, while the sidereal space, together with all of the bodies it contains except for the earth, performs all motions contrary and equal to those that our planet would perform

[51]Note that, in his original translation, Finocchiaro (1989) failed to recognize the difference between the singular/plural meaning of *apparenza/e* (i.e., the common and the technical levels of the concept of appearance/s).

[52]Boscovich (1746, §13) remarked that Bradley's aberration can be explained within this scheme. Perhaps he had slowly recognized that, after all, other explications he might have thought of in 1742 were too complicated to actually be feasible, and their fate was confined to an elegant astronomical exercise. He now seemed convinced that "the propagation of light [with finite velocity] seems to pose *prima facie* the greatest difficulty to the theory of a stationary Earth. For either the propagation of light from point to point should be denied—however, it is proved by the phenomena of the satellites of Jupiter and Bradley's annual aberration of the fixed stars ...—or an inconvenient nutation of all orbits should be admitted .... But finally, in the following manner, every difficulty can be removed. That annual motion together with the Sun and the daily motion around the axis of the equator, which, in the conception of the resting Earth, must necessarily be supposed as common to all celestial bodies, are supposed as common to the light particles as well .... So, those translations of all the bodies and the motions will become more general, and that motion utterly accomplished yearly will cause, in the eye, the same motions that, in the conception of the moving Earth, the light would suffer by the propagation in straight, unalterable line—that is, [it will cause] the impression of a slanting propagation with respect to the eye, hence Bradley's aberration, which therefore proceeds from here."

in Newton's theory. Then, according to Boscovich, "in this system, we say that every phenomena must take place absolutely in the same manner as in the Newtonian system of the moving Earth."

In the subsequent paragraphs (§§ 70–103), various aspects of his solution are examined in order to show that the two systems can be translated into each other. To begin with, Boscovich (1747, § 70) advanced the idea that inertia is only relative to the sidereal "moving" space, which, in turn, would perform two kinds of motion: (1) a daily rotation East to West around the equatorial axis, while the axis itself rotates very slowly around the axis of the ecliptic and oscillates slightly (nutates); and (2) an annual "translation" (*translatio*) on a quasi-elliptic path, in the same manner that the earth travels an approximately elliptic orbit around the sun (the inequalities being due to the action of mutual gravity within the planetary system).

Because this second motion may appear rather unclear, Boscovich (1747, § 71) constructed a model so that it could be better understood. Let us imagine three independent homocentric spheres. The most external one will be completely motionless: Boscovich referred to it as the *Firmamentum*. Let a second sphere within it rotate around its equatorial axis with diurnal motion; it will be called the *Primum mobile*. Inscribed in the latter and concentric to both, there is a third sphere—the *Caelum sydereum*—whose center approximately describes an ellipse with equal and contrary motion to that performed by the center of the earth in Newtonian mechanics. Moreover, let the *Caelum sydereum* have inertia relative to the second sphere (therefore taking part in the diurnal motion), as well as a certain nutation. It is this third sphere, Boscovich assumed, that contains all stars, planets, comets, etc. Now, according to Boscovich, their reference frame is the *Caelum sydereum* and not the *Firmamentum*. In other words, the observable Universe has inertia relative to the moving sidereal space, while the earth is at rest relative to the "firmament"—the latter being only the "infinite, motionless imaginary space," where the motion of a "finite, purely imaginary space" takes place.

Of course, in such a system, the common center of gravity, such as that of the earth and the sun, may be at rest relative to the sidereal space. However, it is kept in motion, relative to the earth, by the moving sidereal space. Boscovich (1747, § 75) instantiated this feature with another model. Let us imagine a vessel at rest on the sea, upon which a passenger is moving in circles around a certain point that is fixed. At the same time, other passengers perform slightly different circular motions (e.g., with different velocities) around the same point, while preserving their mutual distances. At a certain instant, the vessel begins to move with an equal (i.e., rotational) and contrary motion to the first passenger. Of course, due to inertia, the vessel brings along every object and passenger that it contains in its motion. They will preserve their mutual position and "will not suffer any change nor the translation of the vessel. However, the [passenger] who was moving with real motion when the vessel was at rest, is now stationary, since the vessel is moving with motion equal and contrary to him." The passenger is stationary in the sense that he continues to occupy the same position with respect to the water as he had previously, but he is moving with respect to the vessel. In this sense, "the action by which he produced and preserved his own motion when the vessel was at rest will remain when the vessel moves; but . . . it only prevents [the passenger's] translation together with the vessel."

Earth is more or less in the same situation, as Boscovich suggested:

> The same happens to the earth in our system. The sidereal space is like the vessel. If that space were to be at rest for a while, we will have Newton's hypothesis of the moving earth . . . and the explication of so many natural phenomena will be reduced to the very simple law of universal gravitation. Let that space begin to move with motion equal and contrary to all those motions [of the earth], and let every body contained in that space accompany its motion. Certainly, none of the bodies contained in such a space can feel any change; anyone on the earth will observe every phenomenon exactly in the same manner, will measure all particles of matter in the same position relative to himself, as well as to each other. And yet, the earth is at rest, as it would appear to anyone who was to observe it as soon as he was removed from that space. (*ibid.*, § 76)

For Boscovich, this means that there is no objective crucial experiment for determining the state of motion of the earth. In other words, there is no preferred frame of reference within the system. Of course, the bodies in the vessel, which follow a circular path relative to the water (assumed to be stationary), will suffer a centrifugal force and "will endeavor to escape along a straight line" (*ibid.*, § 77). However, in the system of sidereal space, it is exactly this space that is in 'rotational' motion (relative to the firmament, or God's eye). All bodies that are contained in it are assumed to be inertially bound to it; if no other forces act upon them, they will continue the inertial 'rotational' motion that they have in that space. In an attempt to visualize this feature, Boscovich (*ibid.*, § 77) imagined that "the bodies in the vessel are connected to each other and to the vessel itself through rigid bars [*per solidas regulas*] of iron or wood. The rigidity would constrain them to follow all motions of the vessel and preserve their respective positions relative to it," even if "our force of inertia . . . allows all other forces that produce relative motions to freely act within the space."[53]

---

[53]The idea of the sidereal space was resumed some years later in Boscovich's supplements to Benedict Stay's Newtonian poem. He used a similar line of thinking in arguing against Newton's *Gedankenexperiment* of the rotating bucket, where he concluded with a significant statement: "Therefore, it seems to me as evident as possible that we cannot distinguish for any reason whatsoever between absolute and relative motion" (Boscovich 1755b, Suppl. VIII, 68–70, 295); for a comment, see D'Agostino (1989). In the supplement "*De vi inertiae*", the dissertation *De maris aestu* (Boscovich 1747) is explicitly mentioned and Boscovich repeats his argument about the relativity of inertia to a space moving in equal and contrary motions to those of the earth (see Boscovich 1755b, Suppl. XIII, esp. 129–132, 312–314). Finally, in *Theoria philosophiae naturalis*, Boscovich (1763, § 8, footnote *a*) reprised his vision once again: "If this space is at rest, I do not differ from other philosophers with regard to the matter in question; but if perchance space itself moves in some way or other, what motion ought these points of matter to comply with owing to this kind of propensity? In that case, this force of inertia that I postulate is not absolute, but relative; as indeed I explained both in the dissertation *De maris aestu* and also in the Supplements to Stay, Book I, section 13 [see the reference given above]. Here will also be found the conclusions at which I arrived with regard to relative inertia of this sort, and the arguments by which I think it is proved that it is impossible to show that it is generally absolute." In light of this, the theory of sidereal space expressed in Boscovich (1747) and its epistemological implications should be viewed as his mature position.

The only objections that Boscovich regarded as feasible are unphysical, or rather *metaphysical*. One can contend that "the force of inertia the bodies are endowed with is not relative nor bound [*alligatam*] to a certain space, but absolute, that is, a body placed in the void must move along a straight line" (Boscovich 1747, § 84). Or one can reject the translation of the space with equal and contrary motion to those of the earth because it is arbitrary and *ad hoc*. In both cases, Boscovich recognized that these are arbitrary assumptions, but the opposite assumptions ("absolute" inertia and immoveable sidereal space, i.e., Newton's absolute space) are no more and no less arbitrary than his own.

In any case, he concluded, "the entire question [*quaestio*] of the state of rest or motion of the earth" is reduced "to a metaphysical question about the nature of the force of inertia and the state of rest or translation of a certain space" (*ibid.*, § 103). Because they are *metaphysical* questions, they cannot be decided within the limits of astronomy as a physico-mathematical discipline. It also means that people who, like Boscovich, pursue the physico-mathematical sciences can simply put aside such questions and leave them to the philosophers. After all, astronomers can "relatively" be Newtonians in their sidereal space, which allows for a stationary earth. Whether the earth or the space is actually moving is irrelevant, because both frames of reference obey the same physico-mathematical laws: "Every physical effect, every phenomenon in the Universe must occur exactly in the same manner in both conceptions."

* * *

All of this leads to a remarkable epistemological consequence. Mathematics overcomes discrepancies between divergent interpretations because its descriptions only refer to the surface of the things, or their face value. As Boscovich (1747, § 103) concluded, "[in regard to] the improvement of the physico-mathematical disciplines, it does not matter which of them is used." However, this does not exclude the possibility that the choice depends on other factors—e.g., in the case at hand, on religious tenets.

These are, however, different planes. The first plane pertains to the phenomena as they appear to us. Looking from the shoulders of Clavius and Gottignies, for Boscovich, this is the plane of the *physico-mathesis*: in this case, our descriptions have to be Newtonian (or an improvement of Newton's original mathematical descriptions), since they save the phenomena better than any other account. The second plane, instead, pertains to things as they *really* are, namely, things as they appear to the Creator's privileged viewpoint. On this plane, physico-mathematical methods fall short and choice is a matter of faith, metaphysics, or anything else.

There is no contradiction between the two planes, because it can be mathematically proven that our point of view and that of the Creator are compatible (which was the aim of Boscovich's sidereal space). They produce the same physico-mathematical descriptions, and this, in turn, ensures the status of neutrality (1) to mathematics, because it can be tailored to both (or more) ultimate realities and (2) to the mathematician, in terms of the faculty of being agnostic about them.

Chapter 2
God's in His Heaven—All's Right with the World

2.1 A Force Called Inertia and Other Determinations

In the previous chapter, we saw that Boscovich's compatibilist attitude helped him formulate an agnostic mathematical epistemology regarding an ultimate reality as important as the true kinematic state of the earth. For him, astronomy was a seminal ground, because his physico-mathematical attitude first developed in that field. However, astronomy did not suffice to solve the problem of the cosmological system. This could only be solved as far as it could be reduced to the problem of the nature of inertia—a notion that pertains to mechanics. In fact, this was a subject that he had been dealing with since 1740, more or less in the same period as his first compatibilist efforts in astronomy. In turn, the latter might have nourished and empowered his early studies in a physico-mathematical style, but it is in the mechanical works that he developed the agnostic-mathematically neutral epistemology that also factored into the 1747 dissertation on sea-tides. Here, he recognized that, "as far as the force of inertia is concerned—that is, the determination that bodies have to persist in their state of rest or uniform rectilinear motion—its principle is known to us only from observations and experiments and, as we are used to saying, *a posteriori*. For it cannot be absolutely known *a priori*, since the nature of bodies is not known to us" (Boscovich 1747, § 85). But the prototype of this idea, as well as of the discussion that follows in paragraphs 86–89 (including a reference to Euler), were already present in a brief dissertation on the problem of attraction in non-resistant space (Boscovich 1740).[1]

This chapter reprises and develops further my article "Ruggiero Boscovich and 'the Forces Existing in Nature'", published in Science in Context, 30(4): 385–422. https://doi.org/10.1017/S0269889717000266.

[1]Giacomo Zambeccari, a student of the mathematical class at the Collegium Romanum, defended this short treatise in 1740. In that year, Boscovich substituted for Orazio Borgondio as professor of

L. Guzzardi, *Ruggiero Boscovich's Theory of Natural Philosophy*, Science Networks. Historical Studies 60, https://doi.org/10.1007/978-3-030-52093-9_2

The treatise begins with the admission that "practically all scholars of mechanics ascribe to the bodies a certain force by which they preserve their state of rest or uniform rectilinear motion." Note that, from Boscovich's perspective, inertia is literally a force. Of course, this tendency was already present in Newton: Definition III of the *Principia*, as controversial as it is, elucidates what the "inherent force [*vis insita*] of matter" is and embraces both the aspect of inertness as a property of bodies (like extension or impenetrability) and that of a mathematical assumption, "measurable in terms of the induced change of motion."[2] As Boscovich claimed, Newton called such a force "*vim insitam, & vim inertiae*"; in turn, the state that it preserves can only be perturbed "by any other external force" (*ab aliqua vi extrinseca*), such as "gravitation, elasticity, the push of other bodies, and other forces of this kind." On the one hand, Boscovich emphasized, inertia is an assumption:

> [*Vis inertiae*] can be demonstrated neither from the phenomena nor through metaphysical arguments. From the phenomena, indeed, it can never result clearly that gravity, elasticity or other forces of this kind are not intrinsic to the nature of bodies, whose particles either attract or repel reciprocally or, in like manner, are driven to a certain place. Moreover, from the phenomena, it will be much less evident that the motion of bodies cannot decrease by itself ... However, nothing shall ever be achieved from metaphysical arguments if the nature of bodies is not known. And this only unravels to us through the phenomena. (Boscovich 1740, III)

On the other hand, the parallel between inertia as a *vis insita* and the "perturbating" forces as *vires extrinsecae* can be exploited further. Boscovich directly quoted content from the comment to Definition VIII of the *Principia*:

> It seems more convenient [*satius*] to admit that force [i.e., *vis inertiae*] in mechanics exists only in the manner in which Newton admitted attraction, impulse, and propensity in the *Principia*, Lib. I, Def. VIII. This reads as follows: "I use interchangeably and indiscriminately words signifying attraction, impulse, or any sort of propensity toward a center, considering these forces not from a physical but only from a mathematical point of view. Therefore, let the reader beware of thinking that by words of this kind I am anywhere defining a species or mode of action or a physical cause or reason." And Newton had said little above: "This concept is purely mathematical, for I am not now considering the physical causes and sites of forces." (Boscovich 1740, III–IV)[3]

Of course, one might suspect that Newton, in this case, is simply misrepresented. In fact, Boscovich deliberately left out an important passage in the text of the *Principia*. After the statement "I am anywhere defining a species or mode of action or a physical cause or reason [*speciem vel modum actionis causamve aut rationem physicam alicubi definire*]," Newton (1687, 408) added, "or that I am attributing

---

mathematics and, according to a common practice, dissertations were authored by professors and defended by pupils; hence, we may assume that this was Boscovich's own work (this also explains why it is listed as such in the *Catalogus Operum* that closes Boscovich 1763).

[2] I borrow this concept from Gabbey (1971, esp. 36–40; the phrase is quoted from p. 39). On Newton's inclination to consider inertia as a property of matter, see Stein (2002, 283–284, 289) and Janiak (2008, 70–71).

[3] For the quotation from the *Principia*, see Newton (1687, 407).

forces in a true and physical sense to centers (which are mathematical points) if I happen to say that centers attract or that centers have forces [*vel centris (quae sunt puncta mathematica) vires vere & physice tribuere, si forte aut centra trahere, aut vires centrorum esse dixero*]."

The following chart compares Newton's original text with Boscovich's "rendering". The passage that Boscovich extracted from Newton is emphasized (with the original text provided in the footnote):

Newton's text	Boscovich's quote from Newton
Further, it is in this same sense that I call attractions and impulses accelerative and motive. Moreover, *I use interchangeably and indiscriminately words signifying attraction, impulse, or any sort of propensity toward a center, considering these forces not from a physical but only from a mathematical point of view. Therefore, let the reader beware of thinking that by words of this kind I am anywhere defining a species or mode of action or a physical cause or reason*, or that I am attributing forces in a true and physical sense to centers (which are mathematical points) if I happen to say that centers attract or that centers have forces	It seems more convenient to admit that force [of inertia] in mechanics exists only in the manner in which Newton admitted attraction, impulse, and propensity in the Principia, Lib. I, Def. VIII. This reads as follows: "*I use interchangeably and indiscriminately words signifying attraction, impulse, or any sort of propensity toward a center, considering these forces not from a physical but only from a mathematical point of view. Therefore, let the reader beware of thinking that by words of this kind I am anywhere defining a species or mode of action or a physical cause or reason.*" And Newton had said little above: "This concept is purely mathematical, for I am not now considering the physical causes and sites of forces."[4]

One can contend that Newton is only warning the reader that he should not be interpreted as attributing any force to mathematical points at the centers of bodies. Although bodies are real, as Ducheyne (2012, 26) observed, "locating the centripetal force at the centre of a body is a convenient mathematical technique to deal with its overall centripetal force, but it is not to be taken physically, i.e., one should not attribute it 'in a true and physical sense' to centers." Therefore, Newton's claim, if it is read in its entirety, does not necessarily entail that forces *in general* are employed

[4]Newton's text: "*Porro attractiones & impulsus eodem sensu acceleratrices & motrices nomino. Voces autem attractionis, impulsus, vel propensionis cujuscunque in centrum, indifferenter & pro se mutuo promiscue usurpo; has vires non physice sed mathematice tantum considerando. Unde caveat lector, ne per hujusmodi voces cogitet me speciem vel modum actionis causamve aut rationem physicam alicubi definire, vel centris (quae sunt puncta mathematica) vires vere & physice tribuere; si forte aut centra trahere, aut vires centrorum esse dixero*". Boscovich's text with the quotation from Newton underlined: "*Satius videtur eam ipsam vim in Mechanicam admittere eo tantum pacto, quo Newtonus Attractionem, Impulsum, ac Propensionem admisit Princil. I. def. 8. sic enim habet: 'Voces autem attractionis, impulsus, vel propensionis cujuscunque in centrum, indifferenter & pro se mutuo promiscue usurpo; has vires non physice sed mathematice tantum considerando. Unde caveat lector, ne per hujusmodi voces cogitet me speciem vel modum actionis causamve aut rationem physicam alicubi definire': & paulo superius dixerat: 'Mathematicus duntaxat est hic conceptus; nam virium causas, & sedes physicas hic non expendo.'*"

as mathematical tools; after all, they have a physical reality that Book I of the *Principia* was not tasked with investigating.[5]

Now, let us concentrate on Boscovich's application of Newton's passage, no matter how it should be properly interpreted *per se*. As a matter of fact, he dropped the phrase about attributing forces physically to mathematical centers. One can speculate as to whether this was deliberate, or rather that it signaled that those two lines were not particularly significant to him; the crucial aspect is that he left them out. Boscovich simply did not care about the potential physical implications, if any, of the above-quoted passage for Newton. He instead focused on other features:

> Consequently [*igitur*], one shall be allowed to assume [*assumere*], in the very idea of a body transmitted by mechanics [*mechanice tradita*], a determination [*determinationem*] to preserve the state of rest or uniform rectilinear motion it once had. Every cause that changes such a state is called an external force [*vim extrinsecam*], i.e., it does not pertain to that idea [of body]. That cause can perhaps be placed either in the nature of the body itself or in a free law promulgated by the Fabricator of Nature himself, when He created the World. If motion decreases by itself in a void, we will consider that void as a resistant medium, and the resistance in fluids, as recognized through experiments, is increased by a double resistance, one from the nature of the decreasing motion, the other from the resistance of the fluid; but the determinations of motions [*motuum determinationes*] in a non-resistant space will not be physical, but mathematical determinations. (Boscovich 1740, IV)

Even the term *determinatio* adumbrates a Newtonian background. In the *Principia*, it denotes a direction associated with a velocity or a velocity change (acceleration). In this sense, Newton employed it, for example, in the scholium to the definitions and in the comment to the second law of motion, as well as in the unpublished preface to the *Principia*, without hinting at the nature of the causes of motion.[6] Therefore, when Boscovich described inertia or the external forces as determinations, he meant that they are computable in terms of directions having intensity, irrespective of their causes.

---

[5]For a reading of this passage in a causal and realist context about forces, see Janiak (2007, 131–133) and Ducheyne (2012, 25–36). I agree with Ducheyne (2012, 26) that "these statements do not imply a refusal to treat causes and real forces," and that "what physically produces gravity is not part of Newton's analysis in the *Principia,*" but it can be and actually is considered by Newton in other works (e.g., in *Opticks*).

[6]See Newton (1687), respectively, 414 (original Latin text: "*determinatio motus*"/Cohen's translation: "the direction of the motion"), 416 (original Latin text: "*secundum utriusque determinationem componitur*"/Cohen's translation: "according to the directions of both motions"), and 52 ("In the first two books I dealt with forces in general, and if they tend toward some center, whether unmoving or moving, I called them centripetal (by a general name), not inquiring into the causes or species of the forces, but considering only their quantities, directions [lit. determinations], and effects"). Of course, I am not claiming that *determinatio* is *only* a Newtonian concept, but rather that the Newtonian context is more relevant and documentable, as far as Boscovich is concerned. A notion of *determinatio* was employed by Descartes (in the *Principia philosophiae*, II, 44) and possibly exerted an influence on Newton; a similar concept was also used by Leibniz in his manuscript *De Affectibus*. Of course, this issue has raised much interest and discussion: see, for example, the classical paper by Knudsen and Pedersen (1969). See also Garber (1992, 188–193).

2.2 Being Agnostic About the Causal Power of Powers

The notion of determination recurs many times in Boscovich's early treatises on mechanics. In Boscovich (1743, § 1), a dissertation about the motion of bodies subject to central forces proportional to the inverse square of the distance, *determinatio* seems to be applied to inertia alone. *Vis inertiae* is "a certain determination to preserve the state of rest or uniform rectilinear motion in which they are placed"; external forces are measured through the induced change of motion (more precisely, "the change of state they induce in the briefest possible time"). However, if external forces are measured through the motion they induce, this means that only effects matter—thus, intensities associated with the directions. Then, in Boscovich's *De viribus vivis* (1745), the notion of *determinatio* is associated with both inertia and the external forces in a straightforward manner: In this case, inertia is described as "a determination to persist in that state of rest or uniform rectilinear motion once this state has been acquired" (Boscovich 1745, § 10), and an attractive or repulsive force is described as "that determination ... by which the particles tend to approach or separate, whatever the physical cause of this tendency [*conatus*] might be" (Boscovich 1745, § 50). Finally, in the *Theoria*, Boscovich offered a tighter and yet more comprehensive definition:

> As an attribute of [the points of matter], I admit an inherent determination [*determinationem*] to remain in the same state of rest or of uniform rectilinear motion in which they are set once ...; I therefore consider that any two points of matter are determined [*determinari*] to approach one another at some distances, and in an equal degree recede from one another at other distances. This determination I call *force*; in the first case, *attractive*, in the second case, *repulsive*. (Boscovich 1763, §§ 8–9.)

Again, by virtue of the notion of determination, this is associated with causal agnosticism (which does not amount to meaning that there are no causes at all):

> In this determination consists what we have called the force of inertia. Whether it depends upon the free law of the Supreme Creator, or on the nature of points itself, or on some attribute of them, whatever it may be, I do not seek to know; even if I did wish to do so, I see no hope of finding the answer; and I truly think that this also applies to the law of forces ... This term ['force'] does not express a mode of action but the determination itself, whatever its origin. (Boscovich 1763, §§ 8–9)

Let us concentrate on *De viribus vivis*, which signals the crucial step toward his epistemology of force. On the one hand, beyond its intended aim (the discussion of Leibniz's concept of *vis viva*), this dissertation treats mechanical subjects as traditional as falling bodies, collisions laws, and living forces. On the other hand, as previously mentioned, here, Boscovich first presented his idea of a unique force of attraction and repulsion, which he would refine until achieving its final form in the *Theoria*. In a certain sense, the clarification of such basic mechanical notions as velocity and force is preliminary to the introduction of his idea of the unique force as a simultaneous solution to both the problem of collision (for saving continuity in the natural processes) and to the *querelle* of the living forces.

After summarizing Leibniz's argument in his 1686 *Brief Demonstration of a Noticeable Error of Descartes* and the current state of the querelle on the living forces (Boscovich 1745, §§ 3–8) and having announced the main thesis that "there are no living forces in the bodies" (Boscovich 1745, § 19), inertia is introduced as a determination to preserve the state of rest or uniform rectilinear motion. More specifically, "by virtue of this force of inertia, if bodies have null velocity, they rest; if they have some [degree of] velocity, they preserve it as far as a new one is generated by any power [*potentia*]" (Boscovich 1745, § 10).

Since the notion of velocity mattered to Boscovich's definition of inertia, he emphasized a noticeable distinction between two modes of considering velocity, based on the scholastic distinction between *actus primus* and *actus secundus*, meaning that the "secondary act" performs an operation proceeding from the "primary act", which is the "effectedness" of something, its intrinsic capacity to produce a certain effect.[7] Applying these categories to velocity, Boscovich (1745, § 11) calls "velocity *in actu secondo* the relation between the space traversed and the time in which it is traversed" (that is, $v = s/t$); velocity *in actu primo* is, instead, "the determination of a body to have such velocity *in actu secundo*, that is, the determination to cover a determined space in a given time. A body that moves with uniform motion conserves such velocity due to the force of inertia" (Boscovich 1745, § 12). Since inertia is the determination to preserve rest or uniform rectilinear motion, and rest is null velocity whereas uniform rectilinear motion is a certain unchanged velocity, Boscovich concluded that velocity *in actu primo* is "indeed nothing but the force of inertia itself, determined by previous dispositions, that is, either from a first state in which the Creator placed matter while he was creating or by the actions of powers [*actionibus potentiarum*] that acted once" (*ibid.*).

However, note that the above-quoted passages also imply the distinction between *forces* and *powers*, which Boscovich clarified as follows:

> By powers [*potentiae*], we mean those causes that, through their actions, change the state of a body. They determine that that body acquires a different velocity *in actu secundo*, so they are said to produce in it a new velocity *in actu primo*. The instantaneous action due to which

[7]The distinction was a general one and used in many contexts. A codified expression can be found in Thomas Aquinas' *Questiones disputatae de potentia*, q.1., a.1: "Act, however, is twofold: namely, *first* act, which is form; and *second* act, which is being operating [*Actus autem est duplex: scilicet* primus, *qui est forma; et* secundo, *qui est operatio*]". Whereas the association of this distinction with gravity is known (e.g., Strazzoni 2019, 366–367; Malara 2020) and played a role even in Galilei's sources (Wallace 1984, 73), I have been able to find only one occurrence in connection with velocity before Boscovich, namely, in Giovanni Battista Baliani's *De motu naturali gravium solidorum et liquidorum*: "It results from the previous Book II that, while moving, a mobile can move by itself until, after a first motion has been given to it, a virtue or force [*virtus, seu vis*] is impressed upon it by which it can be conducted, without the intervention of anything else, with that velocity with which it moved while that virtue was impressed, and thus with uniform motion [*et proinde motu aequabili*]. This virtue is called *impetus*; it will possibly be different from velocity, since *impetus* is a velocity *in actu primo*, just as, in another manner, *impetus* is the cause of velocity" (Baliani 1646, 79). I am indebted to Ivan Malara for drawing my attention to the scholastic background underlying Boscovich's discourse on this matter.

such velocity is thought to be generated is called an active force [*vis activa*]: for us, this is the one and only force; Leibniz, however, calls it dead force. (Boscovich 1745, § 13.)

Let us remember that an external force is a mathematical determination: It is an abstract mathematical quantity expressing the propensity of an object to perform (i.e., initiate or change) a motion in a certain direction. Let us clarify this issue by means of an example. Imagine that we see a ball thrown by a hand. Boscovich would distinguish between the hand itself—the physical agent that would potentially throw the ball and that he would presumably call a *potentia*—and the "active force". This is the very action of throwing (or "the instantaneous action due to which" the velocity of the ball is generated or changed): the force impressed in the instant of throwing. Such a force is not as "palpable" as the hand. It can be measured according to the laws of motion, etc., but it is not a material agent—rather, it is an abstract mathematical quantity (which entails the advantage that we could ignore the hand and substitute it, such as with a mechanical device, and obtain the same action-of-throwing).

Similarly, according to Boscovich, forces are not physical agents but the effect of physical agents called *potentiae* viz. powers. They include (Boscovich 1745, § 13) impenetrability, gravity, elasticity, and cohesion. The action of a physical power is a determination that is a propensity to perform a certain motion. In Boscovich's vocabulary, this is a force, whereas the term power/*potentia* refers to the causal mechanism.[8] However, as in the case of inertia, the nature of the physical powers can only be hypothesized, whereas forces—being mathematical determinations—are actually measured. As in the case of our hand, which we could ignore and substitute with any mechanical device amounting to the same measured action-of-throwing, here, only determinations matter in order to give a mathematical description of the phenomena, regardless of any hypothesis about the causal mechanism: "Whatever the physical seat of such powers might be, provided that they generate in a body the same velocity *in actu primo* ..., we shall have the same phenomena of motion" (Boscovich 1745, § 14).

This applies, e.g., to gravitation. No matter what produces it (the nature of body, God's will, or something else), the mathematical description dispenses with causal talk:

---

[8]The distinction between powers and forces was reprised later, in Supplement XIV to Stay's Newtonian poem: "Having exposed the force of inertia ..., it still remains to expose the nature of the active forces that divert [*perturbant*] the state of motion from such inertia. *I call powers the causes that generate, accelerate, decelerate, or deflect motion. I call forces their actions, although such causes are called forces too, and vice versa*, most of all when we deal with the produced effect, as well as whenever the causes, through a sort of inversion [*translationem*], are considered as effects in the ordinary language [*communi sermone*]. Causes of this kind are impenetrability ..., gravity, magnetism, elasticity, electricity, the cohesion of parts, fermentation, etc. To be sure, they may be inherent to the nature of bodies themselves or added to and separable from them or may depend from the free law of the Creator.... However, to me, all causes of this kind consist [*consistunt*] of a single and very simple law of forces, common to every material point [*materiae puncta*]" (Boscovich 1755b, 300, emphasis added).

> According to the Newtonians, *gravitas* is a determination of the nature of bodies, or rather a free law of God, so that if two bodies are placed at any distance, even in a vacuum, they immediately acquire a determination to approach one another and to get a new velocity *in actu secundo*, which grows as much as the square of the distance is diminished. Imagine knowing [*intelligantur*] that those bodies exist; imagine knowing the force of inertia, due to which they preserve the previous velocity, if no power is acting; imagine that a certain determined distance is known; imagine knowing that a new velocity *in actu primo*, all conditions being determined, has been generated; then, we will know [that] a continuous takeover [*advenire*] of a new velocity [occurs], if we were to know that the conditions themselves are continuously determined. (Boscovich 1745, § 14)

2.3 Being Neutral About Physical Representations

With all of this, in *De viribus vivis*, Boscovich intended to provide a background for the discussion (and attempted destruction) of the concept of living force. I will not give a detailed account of how he moved from the generation of velocity to *vis viva*;[9] rather, I shall emphasize an aspect of his attitude toward the *vis viva* debate that caused the connection between agnosticism about the nature of forces and mathematical neutralism to emerge.

In the years that followed the publication of Leibniz (1686), Cartesians accepted that they would have to face the challenge about the "true measure" of force, and the controversy represented a major issue within mechanics in the first half of the eighteenth century. Its final point is usually considered to be d'Alembert's *Traité de Dynamique*, which discusses the issue in its preface, leading to the conclusion that "the question cannot consist of more than a completely futile metaphysical discussion or a dispute over words unworthy of still occupying philosophers."[10] However, as pointed out by Iltis (1970a, esp. 135–138), only in the revised second edition of 1758 did d'Alembert distinguish the three cases of dead force mdv (a body has a tendency to move with a certain velocity, which is hindered by an obstacle), quantity of motion mv (a body actually moves with uniform velocity), and living force mv^2 (a body moves with a certain velocity, which is consumed by some cause). D'Alembert concluded that physical descriptions may differ, depending on which effect is considered:

> In all these cases, the effect produced by the body is different, but the body itself does not possess anything more in one case than in another; only the action of the cause that has produced the effect is differently applied. In the first case, the effect is reduced to a simple tendency, which is not properly a measure, since no motion is produced; in the second, the

[9] See, on this, Martinović (1993) and the Introduction to Boscovich (1745).

[10] "*Toute la question ne peut plus consister, que dans une discussion Métaphysique très-futile, ou dans une dispute de mots plus indigne encore d'occuper des Philosophes*" (D'Alembert 1743, xxi).

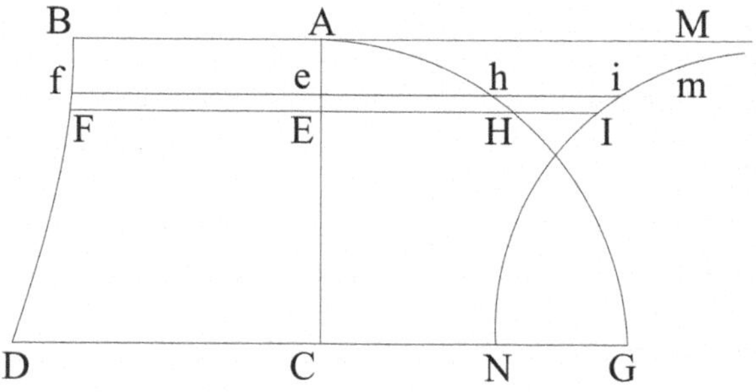

Fig. 2.1 Boscovich's graphical representation of momentum *and* living force as generated by instantaneously acting "pressures." (Modern adaptation from Boscovich 1745, Table I, Fig. 3; courtesy of Andrea Guzzardi)

effect is the space traversed uniformly in the given time, and this effect is proportional to the velocity; in the third case, the effect is the space traversed up to the total extinction of motion, and this effect is as the square of the velocity.[11]

When Boscovich entered the debate with his 1745 treatise, the controversy had already been raging for some decades, but he apparently anticipated some features of d'Alembert's (1758) discussion. Indeed, even Boscovich tended to reduce the controversy to a dispute over words, once the relevant quantities (space, time, and generated velocity) were correctly interpreted.[12] He demonstrated this by means of a twofold diagram (see Fig. 2.1 here below) that separately represented the two measures of force as mv and mv^2.

If we consider powers as acting as time passes, as Boscovich suggested, they "produce a single pressure acting in individual instants of time," so that the pressure they exert "passes into velocity [*transit in velocitatem*] not by multiplication of sort, but just because it is drawn for a continuous time." Boscovich attempts to capture this idea by means of a diagram (Fig. 2.1 above), where pressure (in Boscovich's terms) *EF* is thought to increase over time *AC*. (In modern terms, the diagram plots the pressure variation as a function of time.) However, time can be divided into its infinitesimal elements or *tempuscula*: "If the *spatiolum Ee* is thought of as indefinitely small, because of the indefinitely small difference between the segments *ef* and *EF*, the small area *FEef* is thought of as a rectangle" (Boscovich 1745, § 16). Moreover, due to the infinitesimal difference between two instantaneous pressures in an indefinitely small tempuscule, "in that same tempuscule, even a non-uniform

[11]"*Dans tous ces cas, l'effet produit par le Corps est différent, mais le Corps considéré en lui-même n'a rien de plus dans un cas que dans un autre; seulement l'action de la cause qui produit l'effet est différemment appliquée. Dans le premier cas, l'effet se réduit à une simple tendance, qui n'a point proprement de mesure précise, puisqu'il n'en résulte aucun mouvement; dans le second, l'effet est l'espace parcouru uniformément dans un tems donné, et cet effet est proportionnel à la vitesse; dans le troisiéme, l'effet est l'espace parcouru jusqu'à l'extinction totale du Mouvement, et cet effet est comme le quarré de la vitesse*" (D'Alembert 1758, xxii–xxiii). Note that, in my brief account of the distinctions made by d'Alembert, I used the modern symbolism, which, of course, he did not employ.

[12]On the *vis viva* controversy, see Costabel (1960), Iltis (1967, 1971), Gale (1973), and Papineau (1977). On d'Alembert contribution in particular, see Iltis (1970a). On Boscovich and his *De viribus vivis* in this context, see also Costabel (1961) and Indorato and Nastasi (1993).

accelerated motion is thought of like a uniformly accelerated motion." Now, since pressures cause instantaneous changes in velocity, i.e., they produce accelerations in each tempuscule *Ee*, the relation $v = at$ holds. Or, in the geometrical terms suggested by Boscovich's diagram, the area *BAEF* (instantaneous velocities from *BA* to *EF*, or the increment dv, during time *AE*) represents a velocity.

Therefore, when they are thought about over time, forces are measured through "simple velocities," as opposed to their square, and "the action of powers that generates pressures or velocity—[the action] that we have called the force [of those powers]—is measured by composing the velocities generated in the individual particles, that is, by multiplying the mass for a single velocity." Hence, the final result is momentum as the measure of force when force is viewed as acting over time.

As Boscovich (1745, § 20) acknowledged, Leibniz would agree that this is *a* measure of force. However, he further argued that Leibnizians also allow for a *living* force (*vis viva*), that is, a remnant of powers' actions within the body upon which they have acted. According to Leibniz, it is this force, and not momentum, that preserves its quantum as mv^2. On the contrary, Boscovich contended, this force does not exist. It depends only on the quantities with which one chooses to deal. If, in the diagram, we choose to represent through *AC* the space traversed instead of the time in which an action occurs, and through *EF* Newtonian forces (i.e., such as they are able "to generate, in each tempuscule, velocities proportional to themselves": *ibid.*, § 23; in modern terms, $F = m\frac{\mathrm{d}v}{\mathrm{d}t}$, thus $F \cdot \mathrm{d}t = m \cdot \mathrm{d}v$), then *BAEF* does not correspond to a velocity, but it will be proportional to the square of a velocity. Indeed, force *EF* generates a velocity that is proportional both to *EF* and the time in which *EF* has acted: in modern notation, $dv = EF \cdot dt$. Now, if we assume that, in infinitesimal segment *Ee*, a point moves with uniform linear motion due to the action of force *EF* in the instant of time corresponding to position *E*, then the time corresponding to *Ee* will be proportional to the space traversed and inversely proportional to the velocity: $dt = \frac{Ee}{v}$. From that, we can easily obtain the following by conveniently substituting the terms: $= EF \cdot \frac{Ee}{v}$, and $v \cdot dv = EFEe$, where *EF* is force and *Ee* is (an infinitesimal) space. Finally, applying the infinitesimal method, Boscovich (1745, § 22) concluded that "the square of the velocity of a body falling from *A* (where it was at rest) is proportional to the small area *BAEF*."[13] In other words, force, *as a function of space*, is measured by the square of a velocity. After introducing masses and the modern notation, one obtains this equation: $\int f\mathrm{d}s = \int mv\mathrm{d}v$, with $\mathrm{d}s = v\mathrm{d}t$.

According to Iltis (1970a, 139), Boscovich did not explain why, in the second interpretation of the diagram, time is substituted with space, corresponding to the same segment *AC*, or why he then introduced forces, causing his schema to appear arbitrary. But once Boscovich's agnostic attitude is taken seriously, the reason is easily conceived: both the introduction of forces and the substitution with space *are* indeed arbitrary, because mathematics fits both pictures. The diagram is neutral

[13]Boscovich (1763, § 118 note *f*, § 176 and note *m*) recalled this feature and explained that the areas included between any arc of the curve and the x axis are proportional to velocity squared.

between its possible interpretations, but it can be tailored to both. This has more to do with Boscovich's mathematical approach than with any physical meaning of the geometrical construction itself. After all, he was merely applying the same "mathematical razor" that he had used in the 1740 dissertation on *Motion of bodies thrown through a non-resistant space*.

In the 1745 treatise *De viribus vivis*, this leads to an important consequence as far as living forces are concerned:

> We have shown that nowhere are living forces needed; however, from very easy principles, we can explain all phenomena of motion of any kind whatsoever through an immediate production of velocity by actions of powers that do not leave anything in those bodies other than a different determination of inertia, which persists with inertia itself. If someone, in spite of the uselessness of the living force, yet wants to assume this, he can do what he wishes, phenomena being saved. For, if he would establish that those powers, every time they produce velocity degrees proportional to themselves in individual particles in individual instants of time, also produce proportional forces, then such forces will be as masses multiplied for simple velocities. However, if he desires that, individual equal spaces being given, individual living force degrees occur, which are proportional to the producing forces, then the aggregates of forces will be as masses multiplied by the squares of velocities ... Phenomena—which, as we have seen, only depend upon velocity production—occur in the same manner in both conceptions. ... The first one is that of the Antileibnizians, the other one is typical of the Leibnizians. Because of the uselessness of the living forces, we embrace neither. (Boscovich 1745, §§ 36--37)

Both momenta (mv) and living forces (mv^2) are different descriptions for the same phenomena. This parallels the 1740 paper, in which external forces could "be placed either in the nature of the body itself or in a free law promulgated by the Fabricator of Nature," so they were to be assumed to be mathematical determinations to fit both hypotheses. Yet, in the 1745 dissertation on living forces, neither description "can be demonstrated [as being] false from the phenomena" (*ibid*., § 37), since both are due to mathematical assumptions—let us call them the *temporal* and the *spatial* assumptions—on what (arbitrarily) has to be taken into consideration in each description. So, as Boscovich (*ibid*., § 39) concludes, "whoever does not allow the employment of words in this manner makes a question *de nomine*."

2.4 A Determination of What? Boscovich's Epistemology of Force

The most mature version of this mathematically agnostic, neutral epistemology regarding forces appears in Boscovich's *Theoria* (1758, 1763). As far as his view on inertia and forces as determinations is concerned, the *Theoria philosophiae naturalis* may be considered a radicalization of ideas that Boscovich had expressed in his early works. In establishing the basic elements of his conception, he expressed the following assumptions:

1. The primary elements of matter are non-extended, indivisible and perfectly identical points, which are named *materiae puncta* (Boscovich 1763, § 7). As we shall discuss in Chap. 4, the concept of point of matter was briefly mentioned in the treatise on the living forces (Boscovich 1745, § 61), where the first formulation of the curve appears, but was developed in 1748 in answering a question about the divisibility of matter. This answer represented the core of Boscovich (1757), which is often quoted, commented on, and developed further in Boscovich (1763).[14]
2. The first attribute of these points is "a determination to remain in the same state of rest or of uniform motion in a straight line in which they are initially set" (Boscovich 1763, § 8). This is clearly inertia, and it is introduced and described as it was in his earlier treatises: "In this determination lays what we call *vis inertiae*. Whether this depends on an arbitrary law of the Creator or on the nature of points itself or on some attribute of them, whatever it may be, I do not seek to know; even if I wished to, I see no hope of finding it." Inertia is a description of a kinematic state. To put it as it is in the passage of *De motu corporum projectorum* quoted above (Boscovich 1740, III-IV), inertia is a mathematical assumption in Newton's sense.
3. For any two points, there is a distance-dependent determination to approach or recede from one another: "This determination I call *force* ...; this term does not denote the mode of action, but the determination itself, whatever its origin, whose magnitude changes as the distances change" (Boscovich 1763, § 9)—a qualification reminiscent of Newton's caveat in Definition VIII of the *Principia*. There is nothing special in forces: The term *force* is nothing but a description of a change in kinematic state—an idea that, again, is included in the concept of *determinatio*, as revealed in the previous section.

I argue that the final formulation of Boscovich's force law derives from these three elements. This, of course, shed a somewhat unusual light on the idea that guides Boscovich's curve of forces. James M. Child strongly emphasized point (3) in his introduction to the Latin-English edition of Boscovich's *Theoria* (1763). Based on textual evidence, he remarked that Boscovich's "mutual *vires* are really accelerations, i.e., tendencies for mutual approach or recession of [pairs of] points, depending on the distance between the points at the time under consideration"

[14]Let us remember that Boscovich (1757) was originally composed in 1748. In particular, Boscovich (1757, § 13) contended that the idea of such *material* points is modeled after that of *mathematical* points: "That such points [of matter] can be conceived will not be denied by any geometer, for he conceives mathematical points. As regards extension, points of matter are absolutely similar to them; from them, they differ, because [points of matter] are real, are possibly endowed with real motion, have real properties." Boscovich (1763, § 136) specified those properties: "The first definition of Euclid begins: 'A point is that which has no parts.' After an idea of this sort has been acquired, there is but one difference between a geometrical point and a physical point of matter, i.e., the latter possesses the real properties of the force of inertia and of the active forces that urge the two points to approach towards, or recede from, one another." I will return to this in Chap. 4.

(Child 1922, xiii). Therefore, according to this perspective, there are no forces that actually exist in nature after all.

Following this line of thought, Lancelot L. Whyte has insisted that Boscovich's solution, epitomized in his celebrated graph, only includes aspects that may be called *kinematic*, while dynamics remain out of consideration (see Whyte 1961, 107–108; Child 1922, xii–xiv). In particular, Child (1922, xiii) argued that "this is corroborated by the statement of Boscovich [1763, §§ 118 n. *f*, 176 n. *m*] that the areas under the arcs of his curve are proportional to squares of velocities." We can express such areas analytically as the integral of the acting force multiplied by the space traversed. Hence, we obtain this equation:

$$Subtended\ area = \int f \cdot ds = \int \frac{dv}{dt} \cdot ds = \int v \cdot dv$$

Of course, we are free to think that a particle or a material point accelerates or decelerates because a force moves it. However, we only know (i.e., measure) the effect; therefore, we can abstract from (i.e., be neutral or agnostic about) the cause of motion and simply take the measured value of the force—i.e., the acceleration. A description that does not take the causes of movements into consideration, and only considers the "geometry of motion" given certain initial conditions regarding positions of objects and their relative velocities, is usually called a *kinematic description*.

As we shall see in Chap. 6, Boscovich's analytical expression of the curve only displays spatial terms and parameters, so Child's conclusion appears justified. Notwithstanding all of this, Schofield (1970, 236–237), who sees no ground to emphasize a supposed kinematic approach in Boscovich, "as opposed to [the] dynamic character of [his] theory," has questioned Child's and Whyte's views. After all, Schofield has claimed, by invoking a "determination" of bodies to approach one another, that Boscovich is paraphrasing Newton's "endeavour of bodies to approach each other" (a description of attractive force as it appears in the scholium to Section XI, First Book of the *Principia*), no matter how this is expressed. As such, Schofield concluded, "It seems reasonable, therefore, to say that Boscovich's curve is a curve of attractive and repulsive forces. This is the way he used it and it is the way he was understood in the eighteenth and nineteenth centuries."

It is a historiographically questionable point as to whether the way Boscovich has been understood in the past—for example, by those readers and commentators who experienced his charm in late eighteenth and nineteenth century Britain—should guide a correct interpretation of the significance of his work (I will discuss this issue at length in Chap. 3.4.2). But Schofield is right in recognizing, as he implicitly does, that the distinction between dynamics and kinematics is more a matter of taste than of necessity. The induced change of motion measures the "endeavour of bodies", and *dynamics* means that we are capable of tracing back such "induced change of motion" to the first "induction of change", so to speak.

However, Boscovich denied that a more profound causal level is reachable. (For instance, in our example of the hand throwing a ball, we do not see the hand; we only observe the action of throwing.) We also may call this attitude a "micro-reductive explanatory strategy" (see Čuljak 1995) based on the idea of *determinatio*, though it should not be considered as necessarily antirealist or anticausalist. As we have seen, Boscovich's reader is often reminded that inertia and every external force might be intrinsical to the nature of matter or might depend on God's free will. But this more profound causality cannot be captured by Boscovich's theory, which is only designed to account for forces as determinations. In this sense, such a concept of force is only mathematical: every question about its nature is put aside, and what is relevant is nothing but the measure of the determinations. As such, Boscovich's view is not much different from what de Gandt (1995, 272), in his classical study of Newton's concept of force, dubbed *secularism*: "Scientists can have all sorts of private opinions as to the ultimate realities (or even have none at all), yet a certain common cultural life is possible—with procedures for putting to the proof, and rules governing the confrontation of ideas at a certain level. Mathematics plays a privileged role in the 'neutralization' of the study of force. It is, so to speak, an instrument of 'de-reification.'"

Once the equivalence between forces and determinations has been posed, the counterposition between the kinematic and dynamic interpretations of Boscovich's theory of natural philosophy loses its grounds. Both are part of the same problem—that is, as de Gandt (1995, 12) inquired, "how did force come to be expressed geometrically, and how did mathematics become capable of translating dynamics?" Nevertheless, Child's emphasis on the kinematic aspects of the law can help frame some important consequences of Boscovich's conception. First of all, it makes evident that, according to Boscovich's curve, forces only depend on *geometrical* properties of more or less complicated physical systems. In other words, their positions respect one another. In a very profound sense, they are not properties of matter but are instead properties of distances.

Let us briefly inspect the curve of forces displayed in Fig. 1 of the Introduction and imagine a material point α being in E, where repulsion ceases and attraction toward a hypothetical point fixed in A begins (for the sake of simplicity, we do not consider mutual interactions here). Let us suppose that α tries to move away from A toward G: This motion will be hindered by the determination toward A (i.e., the attraction force "exerted" by our fixed point) in the segment EG; contrarily, if α tries to approach A, this motion will be hindered by the repulsion force dominating in segment AE. Let us now imagine that α has reached the position G: Here, the situation is diametrically opposed. If α tries to proceed to I, its path will be easier because of the repulsive force in GI; if, instead, it attempts to get back to E, the task will also be easier because of the attraction characterizing EG. This suggests that not all of the limit points have the same meaning and, as we shall see, this different meaning is particularly important when the theory of matter comes into consideration (I will come back to this issue in Chap. 4.4).

It is quite obvious that each of these different points may correspond to some "power" in Boscovich's sense. Each one may represent a physically different phenomenon, such as cohesion (for "more stable" positions like E) and some weaker powers like fermentation, elasticity, electricity, magnetism, chemical, or thermal effects for other relatively less stable points. According to Boscovich, we should be agnostic about powers; we only see their "motive" actions. Powers might actually be out there, but they become active (i.e., perform their actions) only at definite distances, in a manner that is described by the geometrical properties of the law of forces, whose empirical details have to be found experimentally. Therefore, phenomena are not the outcome of the interplay of powers that differ according to nature (usually called impenetrability, cohesion, electrical and magnetical forces, etc.); rather, they are the result of *distances*, which give rise to certain determinations.

2.5 God Only Knows

This kind of agnosticism was probably the most important epistemological consequence of Boscovich's education within the Jesuit *physico-mathesis*. For him, empirical measurements and experiments would have been most welcome, as they fit in with its curve of forces. But they would come after mathematical theorization. Experiments would have been welcome as well in order to confirm the theory, but Boscovich did not perform any experiments to this end. He contented himself with noticing that the "nature" of the curve (i.e., the actual forms of interactions between points) and the limit points are to be investigated experimentally (Boscovich 1745, § 56). The third part of the *Theoria* reviewed lots of physical phenomena in qualitative terms—nevertheless, he came across as only being preoccupied with phenomena that fit in with his theory, absent consideration of a quantitative approach. In this sense, his "chemistry" is emblematic:

> Now, those things that are commonly called the elements—Earth, Water, Air, Fire—are nothing else in my theory but different solids and fluids, formed of the same homogeneous points differently arranged; and from the admixture of these with others, other still more compound bodies are produced. Indeed, Earth consists of particles that are not connected together by any force; & these particles acquire solidity when mixed with other particles, as ashes when mixed with oils; or even by some change in their internal arrangement, such as comes about in vitrification . . . Water is a liquid fluid devoid of elasticity . . . Air is an elastic fluid, which in all probability consists of particles of very many different sorts . . . Its particles, however, repel one another with a fairly large force; and this repulsive force of the particles lasts for a long while as the distances are diminished, and pertains to a space that bears a very large ratio to the so much smaller distance, to which it can be reduced by compression. At this distance too, the force still increases, the arc of the curve corresponding to it still receding from the axis . . . Fire is also a highly elastic fluid, which is agitated by the most vigorous internal motions. (Boscovich 1763, § 450)

While, in the case of electricity, Boscovich (1763, § 511) hinted at the studies of Benjamin Franklin and Giovanni Battista Beccaria, in the passage quoted above, he might have had in mind certain experiences or speculations described by Hales (1727), which he knew at least indirectly (Proverbio 2003, 7–12). But Hales was interested in *measuring* force as something palpable and definite enough and as something that experiments were projected to grasp. In Hales' summary ((1727)), one can find, e.g., "Chapter II: Experiments, whereby to find out the force with which Trees imbibe moisture; Chapter III: Experiments, shewing the force of the sap in the Vine in the bleeding season." And again:

> The eleventh experiment shews, with what great force branches imbibe water . . . , where a branch with leaves imbibed much more than a column of 7 feet height of water could in the same time drive thro' 13 inches length of the biggest part of its stem. And in the following experiments we shall find a further proof of their strong imbibing power. (Hales 1727, 78)

This approach became familiar within the "Newtonianizing physics" of the late eighteenth century—above all to those who dealt with small and medium range interactions: electricity, magnetism, heat, chemical processes of any sort, and so on. They were individual forces, different in kind, and could be studied through their particular carriers: electric and magnetic fluids, corpuscles and light particles, phlogiston and caloric, aether, etc. (Heilbron 1979, 67–73). For Boscovich, however, experiments simply measured the approaching and separating of *points* of the same species within *one* homogeneous substance, only more or less rarefied.

Of course, one can observe that neither is Boscovich's aim in the *Theoria* that of a modern physicist nor can his treatise be seen as an attempt at mathematical physics. Nonetheless, what emerges from his "theory of natural philosophy" is a mathematical world picture in which physical assumptions are reduced as much as possible. This explains Boscovich's opposition to conjectures concerning the nature of forces. When he referred to forces, he measured them through velocity changes. One is free to think of them as an effect of "active principles" (such as the cause of gravitation), but Boscovich reasoned that the mathematics governing them only describes the geometry of motion. Is it very difficult to explain the glorious variety of the world by these means? It can be. It is as difficult to explain as, say, the fall of a pin on a desk by applying the Standard Model of particle physics.

As we saw, in mathematics, Boscovich envisioned the possibility of overcoming discrepancies between divergent interpretations. Mathematics allowed for neutrality over the ultimate realities, as well as providing a shared set of procedures for confronting ideas. We are able to construct many different theories of the phenomena, but there is only one mathematical structure unifying them at their basis. Of course, this view applies to cosmology (let us remember the 1747 solution of the sidereal space), but it is something more general, which pertains to natural philosophy as a whole:

> To me, matter is nothing but indivisible points that are non-extended, endowed with a force of inertia, and also mutual forces represented by a simple continuous curve having those definite properties that I stated in Art. 117; these can also be defined by an algebraic equation. Whether this law of forces is an intrinsic property of indivisible points; whether it is something substantial or accidental superadded to them, like the substantial or accidental shapes of the Peripatetics; whether it is an arbitrary law of the Author of Nature, who directs those motions by a law made according to His Will; this I do not seek to find, nor indeed can it be found from the phenomena, which are the same in all these theories. (Boscovich 1763, § 516)

For the rest, those agents, which Boscovich called "powers" (and others called "forces"), might actually be out there—or maybe not. About them, we can have different hypotheses that equally match the phenomena; or—which amounts to the same thing—we can have different hypotheses about the nature of forces (determinations). God only knows. In His Heaven, He may contemplate things as they really are from the instant in which He created them. However, God's Heaven is inaccessible to us. We humans will never be able to transplant God's eye into ourselves, nor will we be able to leave the heaven in which we are confined. We cannot see what things really look like. However, we do have *our* language—i.e., mathematics—to settle the disputes and to overcome difficulties and divergences.

Chapter 3
The Others

3.1 A Matter of Inclinations

Following the line of thought in *De viribus vivis*, I argued that at least one force inversion, at infinitesimal distances, must take place in Boscovich's theory in order to save, e.g., the continuous change of velocity in processes involving the collision of bodies. However, this is only one part of the story. As Boscovich acknowledged as early as 1748, and as early commentators carefully noticed (see Mendelssohn 1759, 354–355; [Montucla] 1760, 59), the idea that forces may invert their sign, passing from being repulsive to being attractive (and *vice versa*) as a consequence of the distances involved, is modelled after a brief passage from the last Query of the *Opticks*, quoted in the second, more theoretical part of an optical treatise, *De lumine* (Boscovich 1748, § 56), as well as in the dissertation presenting his theory of matter, *De materiae divisibilitate* (Boscovich 1757, § 19). The passage reads as follows and its meaning is far from clear: "As in Algebra, where affirmative Quantities vanish and cease, there negative ones begin; so in Mechanicks, where Attraction ceases, there a repulsive Virtue ought to succeed" (Newton 1706, 338; quoted after the English version of Newton 1730, 395). According to Boscovich, Newton's sentence entails the notion that "forces of the same kind [*ejusdem generis vires*] vary according to variated distances, so that they pass from being attractive to being repulsive" (Boscovich 1757, § 19).[1]

[1]Here is the relevant passage from Boscovich (1757, § 19), which includes the quote from Newton: "[He] acknowledges that attraction operates at the smallest distances, and this is the greatest at contact; however, he had this same idea of a transition [*transitus*] from attraction to repulsion according to different distances, and expresses it in the final *Quaestio* of *Optice*, ten pages before the end, with the following words, which offered me the occasion for all these meditations: 'Since Metals dissolved in Acids attract but a small quantity of the Acid, their attractive Force can reach but to a small distance from them. And as in Algebra, where affirmative Quantities vanish and cease, there negative ones begin; so in Mechanicks, where Attraction ceases, there a repulsive

L. Guzzardi, *Ruggiero Boscovich's Theory of Natural Philosophy*, Science Networks. Historical Studies 60, https://doi.org/10.1007/978-3-030-52093-9_3

Of course, this interpretation is somewhat arbitrary. After all, Newton's statement does not imply that we are faced with just *one kind* of force. He conjectured that, where attraction "ceases" (*desinit*), repulsion ought to succeed (*succedere debet*), but here the Latin verb *succedo* simply suggests that repulsion *comes after* or *takes the place of* an attraction as soon as the latter has ceased to act. But why should they *be* the same force with opposite signs? On the contrary, it might well be that attraction and repulsion are forces of two different species.

Boscovich, however, did not seem preoccupied with this kind of philological question. After all, his notion of force as determination dispensed with hypotheses on the possible nature of forces: "We do not employ the name of attractive and repulsive forces in order to mean any physical action at a distance among particles but to express with those words the determination ... by which particles tend to approach each other or to distance [themselves] from each other, no matter what the physical cause of this tendency might be" (Boscovich 1745, § 50). What matters is not but the "determination [of bodies, particles, points...] to approach one another at some distances, and in an equal degree recede from one another at other distances" (Boscovich 1763, § 9). No matter the source from which a force derives, since forces only mean determinations to approach or retreat, there are no reasons to maintain that differently directed forces (determinations) are forces that are different in kind. Hence, this may have led him quite naturally to conceive Newton's original idea as a corroboration of his own concept, namely, as a continuous succession of degrees of the same kind of force, from negative to positive, from repulsion to attraction, and *vice versa*—something that was not necessarily implied by Newton's statement.

This particular turn in the interpretation of a short but significant passage from the *Opticks* signals an important difference, which distinguishes Boscovich's conception from other apparently similar natural philosophies that flourished in post-Newton Britain in the same years as the *Theoria* was conceived on the Continent. Adopting an expression by Schofield (1970), I will refer to them as the "dynamic-corpuscular theory of matter"; in particular, I will label any theorization regarding the microscopic and macroscopic structures or properties of bodies as a "theory of matter" and I will use phrases like "dynamical theories of matter" only in the sense of a conception that gained popularity, most of all among the British Newtonians of the late eighteenth century. According to this, the properties of matter are the effect of the interplay between attractive and repulsive forces emanating from its ultimate

Virtue ought to succeed.' [*'Quandoquidem metalla in acidis dissoluta parvam solummodo acidi portionem ad se attrahunt, liquet vim eorum attrahentem, nonnisi ad parva circum intervalla pertingere. Et sicuti in Algebra ubi Quantitates affirmativae evanescunt, & desinunt, ibi negativæ incipiunt; ita in Mechanicis, ubi Attractio desinit, ibi vis repellens succedere debet.*'] These were Newton's words, which we quoted to make clear that we ascribe to absolutely indivisible points ... forces of the same kind, which vary according to variated distances, so that they pass from being attractive to being repulsive. [*Haec Newtonus; quae adduximus, ut pateret ejusdem generis vires pro mutata distantia variatas ita, ut ex attractivis etiam abeant in repulsivas ... a nobis tribui punctis illis prorsus indivisibilis.*]" As mentioned, the same quotation is also featured in Boscovich (1748, § 56). An indirect reference to this passage from the *Optice/Opticks* appears in Boscovich (1755a, § 168) and in the dedicatory epistle of the *Theoria* (Boscovich 1763, 8–9).

components (no matter if they are called particles, atoms, material points, or something else), which, in turn, are conceived as centers of such forces. I contend that Boscovich can hardly be viewed as a representative of theories of this kind—even if his conception of matter can be regarded as a peculiar Newtonian-styled theory of matter, largely modeled after his acquaintance with the *Opticks*.

This somewhat flies in the face of widespread historical reconstructions, which have connected such currents of thought with Boscovich's natural philosophy. In two cases, Boscovich became acquainted with their authors during his journey to England in 1760, so a direct or mutual influence has also been argued. However, based on chronological and documentary evidence, I will show that Boscovich and the Newtonian dynamicists developed their views independently. Moreover, I will contend that, whereas the inclination to theorize about forces and matter was a shared, distinguishing trait of the Newtonians, Boscovich's own theory of matter, which I will present in this chapter, simply dispensed with such theorizations. But, of course, the success of the dynamical theories of matter in Britain can explain why he was considered a supporter of these views and was seen as a forerunner of Faraday's field-theoretic conception.

3.2 Newton's Ambiguity Disentangled

Scholars have sometimes remarked[2] that the early Newtonian expositors did not take Newton's apologetics about the usage of forces in the *Principia* too seriously—e.g., the exhortation mentioned in the previous chapter, to not think "by words of this kind [attraction, impulse, force, etc.] I am anywhere defining a species or mode of action or a physical cause or reason." Of course, they have had good reason to do so. The author of the *Principia* often declined to speculate about the causes, seat, and mechanism of, say, gravity; however, some of his formulations are more oscillating and ambiguous than he would have been willing to admit.[3] In fact, he famously launched an intriguing, as well as obscure, conjecture about gravity in an often-quoted letter addressed to Richard Bentley in 1693: "It is unconceivable that inanimate brute Matter should, without Mediation of something else, which is not material, operate upon and affect other Matter without mutual Contact." He also confessed to considering it as "so great an absurdity" that gravity should be assumed as "innate, inherent, and essential to matter, so that one body may act upon another at a distance through a vacuum, without the mediation of anything else." In the end,

[2]See, e.g., Schofield (1970, 8–14), Heilbron (1979, 47–58).

[3]Heilbron (1979, 48–49) exemplified this ambiguity with Proposition VII of Book III: "It argues that 'all parts of any planet A gravitate towards any other planet B', a formulation which, as Newton's editor Roger Cotes told him, seems to imply the hypothesis that the power of gravitating resides *in* the several parts of matter" (Italics in the original text). See also McMullin (1978, 52–53, 57–74) and Cohen's comment in Newton (1687, 277–280). For a discussion of the "reality of forces" in Newton, see Janiak (2007, 2008, 50–86) and Ducheyne (2012, 25–45; 2014).

Newton might have actually ascribed the "action at a distance" of gravity to a more or less complicated mechanism of a medium, such as, perhaps, a non-mechanical aether permeating all matter.[4]

The Newton-Bentley correspondence was not published until 1756, but it is reasonable to assume that the general arguments began to circulate very soon after the exchange took place. In any case, those who were suspect of Newton's speculations when speaking of force were further powered by the publication of the *Opticks* with its final guesses, and the Queries, growing in number from the first edition (1704) to the second English edition (1717–1718)[5] and containing speculations about so-called "active principles", e.g. the cause of gravitation, that of fermentation, that of the cohesion of bodies. Without them, Newton argued in Query 31, matter would be confined to inertness.[6]

For the early Newtonians, all of this certified Newton's strong belief in the existence of forces and allowed for the prospect of a scientific inquiry of their causes by means of experiments. Whatever intention Newton might have had with the Queries, for the Newtonians, they represented an urgent call for research, which they could unequivocally determine in the last phrase of the *Opticks* before the Queries began: "Since I have not finish'd this part of my Design, I shall conclude, with proposing only some Queries *in order to a farther search to be made by others*" (Newton 1730, 338–339; italics added). Moreover, they interpreted this mandate by taking on the task of assigning mathematical form and empirical magnitude to any of the various forces of attraction and repulsion that were emerging in the investigation of the short-range and medium-range interactions.

[4]See Schofield (1970, 13–14), Janiak's Introduction in Newton (2004, esp. xxiii–xxiv), and Janiak (2008, 99–102).

[5]See Schofield (1970, 10–11), Heilbron (1979, 51–55), and Hall (1993, 127–162, 238).

[6]"Seeing therefore the variety of Motion which we find in the World is always decreasing, there is a necessity of conserving and recruiting it by active Principles, such as are the cause of Gravity, by which Planets and Comets keep their Motions in their Orbs, and Bodies acquire great Motion in falling; and the cause of Fermentation, by which the Heart and Blood of Animals are kept in perpetual Motion and Heat; the inward Parts of the Earth are constantly warm'd, and in some places grow very hot; Bodies burn and shine, Mountains take fire, the Caverns of the Earth are blown up, and the Sun continues violently hot and lucid, and warms all things by his Light. For we meet with very little Motion in the World, besides what is owing to these active Principles. And if it were not for these Principles the Bodies of the Earth, Planets, Comets, Sun, and all things in them would grow cold and freeze, and become inactive Masses; and all Putrefaction, Generation, Vegetation and Life would cease, and the Planets and Comets would not remain in their Orbs .... It seems to me farther, that these Particles have not only a *Vis inertiæ*, accompanied with such passive Laws of Motion as naturally result from that Force, but also that they are moved by certain active Principles, such as is that of Gravity, and that which causes Fermentation, and the Cohesion of Bodies" (Newton 1730, 375–376).

3.2.1 *Hales' Amphibious Air*

Particularly in the last Query (31), Newton had suggested that the forces of attraction (such as cohesion in the short range and gravitation on the largest scale) could not exhaust all of the phenomena of nature. Some of them—for example, those that take place in the air—could only be accounted for as the interplay of attractions and repulsions.[7] However, repulsion did not experience immediate success among the Newtonians.[8] Stephen Hales, in his *Vegetable Staticks* (1727), was probably the first to make use of it in a systematic and extensive manner. In order to describe the phenomena of vaporization, he considered air particles as being endowed with elasticity—i.e., with a repulsive force that can be reversed into attractive, and *vice versa*, under appropriate conditions. Air would possess the "amphibious property" of being potentially "changed from a strongly repelling to as strongly an attracting state and back again" (Hales 1727, V–VI).[9] Two different states are distinguished: *elastic air*, by which the particles tend to distance themselves from one another (so that they display a repulsive power), and *fixed air*, whereby the elasticity gets lost and repulsion between particles is superseded by attraction. Claiming that he was following Newton's *Opticks* (see, in particular, Newton 1730, 484–485), Hales ascribed the generation of a fixed, attractive state of the air—i.e., the destruction of elasticity—to processes of combustion and vaporization (or sublimation) in which "sulphureous bodies" take part (Hales 1727, esp. VI, 297–298).[10]

---

[7]See Thackray (1970, 32–38), Schofield (1970, 11–18), Heilbron (1979, 54–55), Quinn (1982, 110–112), and Hall (1993, 142–147).

[8]The very early expositors of Newton sparingly used repulsions in their accounts of short-range interactions. See Heilbron (1979, 63–64), and Quinn (1982, 112–116).

[9]In particular, according to Hales (1727, 243–244), "It is evident from the foregoing Experiments on respiration, that some of the elasticity of the air, which is inspired, is destroyed; and that chiefly among the vesicles, where it is most loaded with vapours; whence probably some of it, together with the acid spirits, with which the air abounds, are conveyed to the blood, which we see is by an admirable contrivance there spread into a vast expanse, commensurate to a very large surface of air, from which it is parted by very thin partitions; so very thin, as thereby probably to admit the blood and air particles (which are there continually changing from an elastick to a strongly attracting state) within the reach of each other's attraction, whereby a continued succession of fresh air may be absorbed by the blood. And in the analysis of the blood, either by fire or fermentation in Exper. 49 and 80, we find good plenty of panicles ready to resume the elastick quality of air: But whether any of these air particles enter the blood by the lungs, is not easie to determine; because there is certainly great store of air in the food of animals, whether it be vegetable or animal food. Yet when we consider how much air continually loses its elasticity in the lungs, which seem purposely framed into innumerable minute meanders, that they may thereby the better seize, and bind that volatile Hermes [i.e.. mercury]: It makes it very probable that those particles which are now changed from an elaftick repulsive, to a strongly attracting state, may easily be attracted thro' the thin partition of the vesicles, by the sulphureous particles which abound in the blood."

[10]Hales' model is extremely rich and embraces various phenomena, such as distillation, fermentation, and combustion, though it is not my aim to present it in this book. For extensive accounts, see Schofield (1970, 74–79), Allan and Schofield (1980, 30–64), and Quinn (1982, 116–119).

As Quinn (1982, 111 and 118) observed, this aspect of Hales' theory of the air, in particular, signals a remarkable turn in the original Newtonian scheme of attraction and repulsion. In the last Query, Newton (1730, 375) ascribed to "the small Particles of Bodies certain Powers, Virtues or Forces, by which they act at a distance," here leaving aside the problem concerning whether such powers, virtues or forces would be shared equally by all of the particles or if only some of them were endowed with certain special powers. To present it as a dilemma, can all corpuscles exert both attractive and repulsive forces, or is there a sort of "specialization" and do forces depend on the constitution of the corpuscles? Newton tended to regard matter as homogeneous with respect to forces, in the sense that all particles could, in principle, be affected by, or be endowed with, both attraction and repulsion. Of course, specific forces (except gravity, which is universal) act selectively: they affect different particles differently.[11] However, as pointed out by Quinn (1982, 118), "for Newton, the particles of ordinary matter can be elastic like air; although, as vapors, their repulsive force is much weaker than that of permanent air."[12]

Hales was by far more radical on this. According to him, *air* is amphibious—not the other matter. Only air can be both attractive and repulsive, whereas ordinary matter is always attractive. As such, Hales' distinction between the two principal states of the air entails a distinction between two kinds of matter: elastic-inelastic corpuscles (air) and permanent inelastic corpuscles (ordinary matter). A dynamic balance of these qualitatively different species is necessary so that the whole Universe does not collapse in upon itself. On the one hand, "there is good reason from these Experiments to attribute the fixing of the elastick particles of the air to the

[11]See, for example, Newton (1962, 324–325): "But just as a magnet is endowed with a twofold force, both the force of gravity and the magnetic force, so there can be various forces of the particles [of many substances], which descend from various causes. Different must be the force of the particles of oil that causes them to mutually attract, and different must be that by which they repel the particles of water. Particles of metal attract particles of solvent acids, but particles composed by both substances floating in water repel one another." [*"Quemadmodum vero Magnes vi duplici praedita est, altera gravitatis & altera magnetica; sic earundem particularum variae possunt esse vires ex varijs causis oriundae. Alia debet esse vis particularum olei qua se mutuo attrahunt, et alia qua fugiunt particulas aquae. Particulae metallorum attrahunt particulas acidorum solventium, at particulae ex utrisque compositae in aqua natando se mutuo fugiunt."*] I am indebted to Niccolò Guicciardini for drawing my attention to this passage from Newton's manuscripts.

[12]For the present purpose, it suffices to cite more extensively the above-quoted passage from the *Opticks*: "Since Metals dissolved in Acids attract but a small quantity of the Acid, their attractive Force can reach but to a small distance from them. And as in Algebra, where affirmative Quantities vanish and cease, there negative ones begin; so in Mechanicks, where Attraction ceases, there a repulsive Virtue ought to succeed. And that there is such a Virtue, seems to follow from the Reflexions and Inflexions of the Rays of Light. For the Rays are repelled by Bodies in both these Cases, without the immediate Contact of the reflecting or inflecting Body. It seems also to follow from the Emission of Light; the Ray so soon as it is shaken off from a shining Body by the vibrating Motion of the Parts of the Body, and gets beyond the reach of Attraction, being driven away with exceeding great Velocity. For that Force which is sufficient to turn it back in Reflexion, may be sufficient to emit it. It seems also to follow from the Production of Air and Vapour ..." (Newton 1706, 338; quoted after the English version of Newton 1730, 395).

strong attraction of the sulphureous particles with which [Newton] says it's probable that all bodies abound more or less" (Hales 1727, 298). On the other hand, air also permeates every single compounded body:

> Thus upon the whole, we see that air abounds in animal, vegetable and mineral substances; in all which it bears a considerable part; if all the parts of matter were only endued with a strongly attracting power, whole nature would then immediately become one unactive cohering lump; wherefore it was absolutely necessary, in order to the actuating and enlivening this vast mass of attracting matter, that there should be everywhere intermixed with it a due proportion of strongly repelling elastick particles, which might enliven the whole mass, by the incessant action between them and the attracting particles: And since these elastick particles are continually in great abundance reduced by the power of the strong attracters, from an elasltick, to a fixt state; it was therefore necessary that these particles should be endued with a property of resuming their elastick state, whenever they were disengaged from that mass, in which they were fixed. (Hales 1727, 313–314)

The microscopic structure of matter governs the forces. It *causes* the macro-level to appear in its colorful dress of well-structured bodies approaching and distancing themselves from one another. In turn, experimentally measuring the forces refers to the possibility of inquiring into their causes—i.e., paraphrasing Newton's design as advanced in the incipit of the *Opticks*, to propose and prove *the properties of matter* based on reason and experiments.[13]

3.2.2 The Spheres of Activity of John T. Desaguliers and John Rowning

Hales' work was an immediate success among his contemporary comrades in the Newtonian army, and it had a considerable influence on the successive generations of researchers, both in Britain and in continental Europe.[14] One of his first and, in turn, most influential brothers-in-arms was Rev. John Theophilus Desaguliers, the Huguenot refugee, successor of John Keill at Oxford as lecturer in natural philosophy, and then demonstrator of the Royal Society in London.[15] As a curator of experiments (having succeeded Francis Hauksbee in 1713), the Society solicited from Desaguliers a *résumé* of *Vegetable Staticks* when, at the meeting on April 13, 1727, Hales presented the new work. Desaguliers published an enthusiastic review and a summary in the *Philosophical Transactions* and, in subsequent

[13]"My Design in this Book is not to explain the Properties of Light by Hypotheses, but to propose and prove them by Reason and Experiments: In order to which, I shall premise the following Definitions and Axioms" (Newton 1730, 1).

[14]For a comprehensive account of Hales' influence, see Guerlac (1977), Allan and Schofield (1980, 119–140), and, particularly for his influence on Priestley, Schofield (1997, 262–264).

[15]For an overview of Desaguliers' life and works, see Carpenter (2011).

meetings, he offered the members a demonstration of the experiences described in Hales' book.[16]

The study of Hales' work led to a noticeable and long-lasting impression on Desaguliers. As Schofield (1970, 81–82) noted, in Desaguliers' works, the first use of both attractive and repulsive forces traces back to 1728, whereas repulsions do not appear in the experimental papers he had published thus far. The two-volume *Course of Experimental Philosophy* (I vol. 1734; II vol. 1744), which expanded his lecture notes and collected many individual papers published in the *Philosophical Transactions*, contains the below praise, which also clarifies how Desaguliers viewed his own work as following in the footsteps of Newton:

> When Sir Isaac Newton publish'd a second Edition of his *Opticks*, in year 1717, he added Queries to the 3rd Book. Those, together with the rest of the Queries, contain an excellent Body of Philosophy, and upon Examination appear to be true; though our incomparable Philosopher's Modesty made him propose those Things by Way of Queries, which he had Observations enow to satisfy himself were true; he was unwilling to assert any Thing that he could not prove by Mathematical Demonstration or Experiments. This made a great many People consider what he says in them as mere Conjectures; and I know very few, besides the Reverend Learned Dr. Stephen Hales and myself, that look upon them as we do on the rest of his works ... I refer my Reader to the Doctor's excellent Treatise of *Vegetable Staticks* ... That Air is sometimes in a fix'd, and sometimes in an elastick State, (the last one being the only State in which Philosophers took it to be before Sir Isaac Newton) as well as how it may be changed from one of those States to the other, sometimes with great Ease, and sometimes with great Difficulty, has been shewn by Dr. Hales, by a vast Number of curious Experiments. (Desaguliers 1744, 403)

However, according to Desaguliers, the repulsive virtue is not characteristic of air alone. A similar force explains the incompressibility of fluids. The particles of water, he maintained, are also in possession of "a repulsive Quality of immense Force" and "all the Liquors which have Water for their Basis are endowed with this repellent quality"; finally, "even Metals when they are in Fusion, are incompressible by this Property" (Desaguliers 1744, 337). Of course, repulsion alone cannot explain why fluids collect in more or less large volumes or why they form drops. Hence, Desaguliers imagined them as being governed by a mechanism of alternating "spheres of action" of repulsive and attractive forces, with which he detailed what are now commonly referred to as *phase transitions*:

> To this repulsive Force, whose Sphere of Activity extends but a little way (perhaps not beyond the Surface of the constituent Particles of Water) succeeds an attractive Force, that we shall call Attraction of Cohesion, which begins where the other ends, and confines its Extent. It is by this Attraction of Cohesion, which acts in a Sphere, that the Particles of Water join'd together form Drops till a certain Bigness, without the Sphere of Repulsion abovementioned ... When the Particles of Water are separated by any Cause whatever that puts them into motion, the Attraction of Cohesion yields by little and little, and acts no longer at a

[16]Carpenter (2011, 71) reported the circumstance. The review and the summary were respectively published as J.T. Desaguliers, "An Account of a Book entitl'd Vegetable Staticks ... by Stephen Hales B.D. F.R.S. ...", *Philosophical Transactions*, 34 (1726–1727), 264–291, and "The Conclusion of Dr. Desaguliers' Account of Mr. Hales' Vegetable Staticks", *ibid.*, 35 (1727–1728), 323–331.

> distance something sensible; and then a second repellent Force may succeed to the Attraction of Cohesion; and the Particles acquire a Force, (which in this case we shall call centrifugal) by which they repel each other, and fly off even to very great distances, oftentimes taking up 14 thousand times more Space than when they were join'd in Water. This will happen by the Action of that Degree of Heat which makes Water boil; the Water being then changed into a Vapour, whose Elasticity is equal to that of the Air ... Air is a Fluid, whose Parts are endowed with a centrifugal Force, whereby they repel each other without touching ... Every compressible Fluid that loses the centrifugal Force of its Parts, (and consequently its Elasticity) ceases to be a Fluid, and becomes a Solid. (Desaguliers 1744, 337–338)

Thus, in spite of his admiration for Hales, Desaguliers was able to formulate an "extended theory" of attractive and repulsive forces, which was only partly in accordance with that of *Vegetable Staticks.* Of course, Desaguliers' conjectures appeared after Charles Du Fay (1733) had announced, in a letter published in the *Philosophical Transactions*, his observation of two different kinds of electrification: the "vitreous" and the "resinous". As he pointed out, bodies that are electrified with the same kind of electricity (e.g., resinous-resinous) repel one another; bodies that are electrified with electricity of different kinds (e.g., vitreous-resinous) attract one another. To be sure, Desaguliers was aware of this discovery and its importance, so it might well be that, as Quinn (1982, 119–120) pointed out, this led him to recognize that repulsive forces were not confined to the air, as in Hales' theory.[17] In addition, a reference to magnetic experimentations might have played a role.

In fact, Desaguliers (1739, 175) maintained, "Attraction and Repulsion seem to be settled by the Great Creator as first Principles in Nature." Some natural philosophers, he continued, have tried to reduce elasticity to a force of attraction, but this explanation does not seem to cover every phenomenon:

> If [in Fig. 3.1a] the String AB be consider'd as made up of Particles lying over one another in the manner represented at ADB; it is plain, that if the Point D be forcibly brought to C, the Parts will be pull'd from each other; and when the Force, that stretch'd the String, ceases to act, the Attraction of Cohesion (which was hinder'd before) will take place, and bring back the String to its former Length and Situation after several Vibrations. Now, though this

[17]Nevertheless, Desaguliers seemed to have conceived his theory of electricity as an extension of Hales' theory of the air. He concluded his *Dissertation Concerning Electricity* with the Newtonian disclaimer that "I have not endeavour'd to guess at the Cause of Electricity." However, "if Conjectures are desir'd, here follow some: I suppose Particles of pure Air to be Electric Bodies always in a State of Electricity, and that Vitreous Electricity." Then, Desaguliers described an experiment conducted by Hauksbee that, according to him, proved "my [Desaguliers'] Conjecture ..., that *the Air is electrical*". Finally, Hales' theory is interpreted in consideration of such a conjecture: "In the Reverend and Learned Dr. Hales' *Vegetable Staticks*, several of his Experiments shew, that Air is absorb'd, and loses its Elasticity by the Mixture of sulphureous Vapours, so that four Quarts of Air in a Glass Vessel will, by the Mixture of those Effluvia, be reduc'd to three. Will not this Phenomenon be explained by the different Electricity of Sulphur and Air. The Effluvia of Sulphur being electrick repel one another: and the Particles of Air being also electrick, do likewise repel each other. But the Air being electrical of a vitreous Electricity, and Sulphur of a resinous Electricity, the Particles of Air attract those of Sulphur, and the *Moleculae* compounded of them becoming non-electrick lose their repulsive Force" (Desaguliers 1742a, 39–41). These passages are also resumed in Desaguliers (1742b, 140–142), where they are connected with Du Fay's "Assertion of Two Sorts of Electricity."

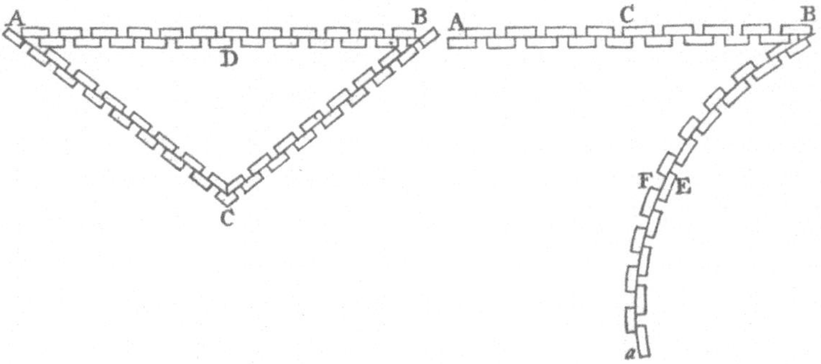

Fig. 3.1 (**a**) (on the left) and (**b**) (on the right). Elasticity of metal bars when viewed as a force of attraction only, according to Desaguliers (1739). The figures are reproduced from *Philosophical Transactions*, 41(454), 1740, Table 1, Figs. 1 and 2

> seems to agree pretty well with the Phaenomena of a String in Motion, it will by no means solve the Elasticity of a Spring fasten'd at one End, and bent either way at the other, like a Knife ox Sword-blade, as in [Fig. 3.1b] For if such a Spring be bent from A to a, the Particles on the Side C, which now becomes convex, will be farther asunder at F, while the Particles at D, carried to the concave Part E, will come closer together: So that the Attraction, instead of making the Spring restore itself, will keep it in the Situation in which it is, as it happens in Bodies that have no Elasticity, where perhaps only Attraction obtains. (*ibid*., 179)

There is, however, a more promising way to explain the same phenomena—i.e., (*ibid.* 180) "to consider both a repulsive and an attractive Property in the Particles, after the manner of the black Sand, which is attracted by the Loadstone, and has been shewn by [Pieter van Musschenbroek] to be nothing else but a great Number of little Loadstones." Perhaps, as Desaguliers insinuated, elastic strings, such as that of steel, are composed of particles that have poles like magnets. In such a case (*ibid*., 181), "may not a Spring of Steel, or other Springs, consist of several Series of such Particles, whose Polarity and Attraction acting at the same time, will shew why such Bodies, when they have been bent, vibrate, and restore themselves?" Based on this, a conjecture can be advanced:

> Let us suppose AB [Fig. 3.2a] to be two little Spheres or component Particles of Steel, in which, at first, we will suppose no Polarity, but only an Attraction of Cohesion. Then, whether the Particles have their Contact at *c, d, e, n* or at *δ, ε, ς*, their Cohesion will be the same; and the least Force imaginable will change their Contact from one of those Points to another ... But if we suppose the Point *n* in each Spherule to be a Pole with a Force to repel all the other Points *n* in any other Spherule, and likewise *ς* another Pole, repelling the other Points *ς*; the Spherules will cohere best, and be at Rest in that Position where the Points *c, c*, are in Contact, and *n* and *ς* at equal Distances on either Side. For if the Spherules be turn'd a little, so as to bring the Points *d, d*, into Contact [Fig. 3.2b] the Poles *n, n*, being brought nearer, act against each other with more Force than the Points *ς, ς*, which are now farther off, and consequently drive back the Spherules to the Contact at *c, c*, beyond which continuing their Motion, they will go to *δ, δ* [Fig. 3.2c], and so backwards and forwards, till at last they rest at *c, c*, which we may call *the Point of Aequilibrium for Rest* in a Spring. (*ibid*., 181–182)

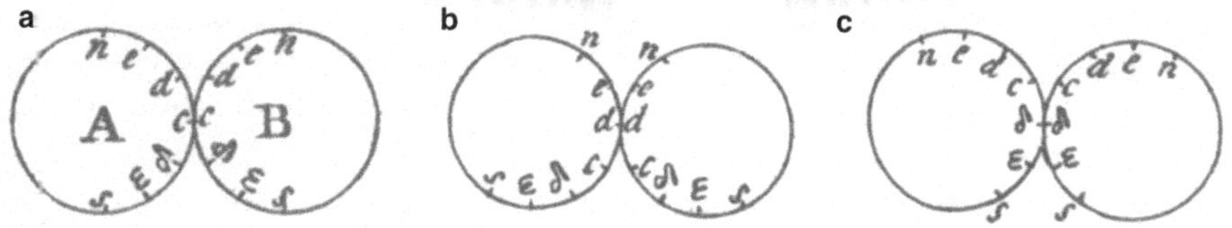

Fig. 3.2 (a–c) (from left to right). Desaguliers' explanation of elasticity in metal bars, in which he presents them as constituted by particles endowed with poles. From *Philosophical Transactions*, 41 (454), 1740, Table 1, Figs. 1 and 2

Hales' image of nature was very different from the idea that transpires from the above-quoted passage. The author of the *Vegetable Staticks* suggested that an underlying dynamic equilibrium made up of particles of attractive-repulsive air twisted with attractive matter was integral to the whole of nature. For Desaguliers, attraction and repulsion are ubiquitous and may be found in the air and elsewhere. However, the fact that repulsion and cohesion are ubiquitous—in Desaguliers' words (*ibid.*), the fact that they "seem to be settled by the Great Creator as first Principles in Nature"—does not necessarily imply that any chunk of matter, or any single particle, should be affected by both. It simply means that, in nature, both actually exist (and, of course, repulsion is not confined to air), and they act in a variety of manners. Repulsion can be found in *some* solids (e.g., elastic steel bars), in *some* liquids (incompressible fluids), and in vapors. Additionally, attraction and repulsion are sometimes properties of matter in a particular physical state, such as the first sphere of activity in fluids, which probably extends "not beyond the surface of the constituent particles of water," or the particles with poles in magnets and elastic solids. However, they are occasionally obtained as a result of mechanical processes. This is the case of the centrifugal force in liquids ("the particles acquire a force, which in this case we shall call centrifugal, by which they repel each other... Every compressible fluid that loses the centrifugal force of its Parts, and consequently its Elasticity, ceases to be a fluid, and becomes a solid").

It is questionable as to whether Desaguliers' explications were really intended to be a generalization of the model of alternating spheres of attraction and repulsion, as some scholars have suggested. In an attempt to associate him with Boscovich, Heilbron (1979, 66) claimed that, according to Desaguliers, "the particles of matters [are] at the centers of alternating spheres of attractive and repulsive force" and their approaching or receding from one another "was an accident of distance." In the same vein, Quinn (1982, 125) argued that he "supposed that every corpuscle is surrounded by spheres of attractive and repulsive force. One corpuscle exerts either attractive or repulsive force upon another depending on their separation. All corpuscular phenomena, Desaguliers claimed, could be treated within his system." However, if we take his explanation of elasticity at face value, we ought to suppose that the *spherules* composing an elastic bar are endowed with poles. *The manner* in which attraction and repulsion take place, instantiated by the above-quoted passage, is, of course, an accident of distance; however, *the reason* why they take place is that each sphere of the elastic material is structured in a certain manner, for it has two contrary poles. We can also assume that the elastic properties of a particular bar may depend on the distribution of the poles. This was a conclusion that Desaguliers (1739, 185) seemed to be willing to draw: "If the Poles that we have consider'd be plac'd quite irregularly, there will be no Elasticity at all." In any event, repulsion and attraction depend on the microscopical structure of matter: one pole will always exert the same action on another pole, depending upon its polarity and independent of the distance (which influences the intensity of the force). Hence, paraphrasing Heilbron, one may conclude that, according to Desaguliers, repulsion and attraction are, in fact, *an accident of substance*.

It is not easy to say if the idea of one or more "spheres of activity" is derived from others or was independently developed by Desaguliers. Some years before his study of the causes of elasticity (1739), John Rowning had developed a similar conception, though not in quantitative terms. As a pupil, then a fellow, and, finally, a rector of the Magdalen College in Cambridge, he authored a large exposition of a Newtonian theory of matter entitled *A Compendious System of Natural Philosophy* (1734–1743).[18] In the book, attraction and repulsion are specifically addressed in Parts I and II. They are "powers or active Principles, probably not essential or necessary to its existence [i.e., of matter], but impressed upon it by the Author of its Being ... Attraction is of two kinds. (1) Cohesion, or that by which the several particles whereof Bodies consist, mutually tend toward each other. (2) Gravitation, or that by which distant Bodies act upon each other" (Rowning 1753, I, 12). Cohesion "act[s] only upon contact or at very small distances," whereas gravitation only acts between distant bodies (*ibid.*, I, 13.)

In regard to repulsion, this "is the property in Bodies, whereby if they are placed just beyond the Sphere of each other's Attraction of Cohesion, they mutually fly from each other" (*ibid.*, I, 17). Rowning does not explain more than this about repulsion. As a means to support this assertion, he referred to chemical experiments with mixtures and to Query 31 of the *Opticks*, where Newton gave "an undeniable Proof of this Repulsive Force." A remarkable sophistication of this idea and a noticeable innovation are expounded in Part II, which deals with hydrostatics and pneumatics:

> Each Particle of a fluid must be surrounded with three spheres of Attraction and Repulsion one within another: the innermost of which is a Sphere of Repulsion, which keeps them from approaching into Contact; the next, a Sphere of Attraction diffused around this of Repulsion, and beginning where this ends, by which the particles are disposed to run together into Drops; the outermost of all, a Sphere of Repulsion whereby they repel each other, when removed out of that Attraction. If this were allowed and we might go on, and suppose the Particles of all Bodies to attract and repel each other *alternately* at different Distances, perhaps we might be able to solve a great many Phaenomena relating to small Bodies, which

[18] As reconstructed by Schofield (1970, 35–39), the book, which exerted some influence, has a complicated and uncertain chronology. It is composed of four parts, each of which was issued separately between 1734 and 1743. The preface should trace back to 1743; Part I, 1734; Part II/1, 1735; Part II/2, 1736; Part III/1–2, second edition, 1743; Part IV, 1742. Starting with 1753, it seems that all parts were finally collected into a two-volume book and published several times in the subsequent years; still, they remained separately paginated.

> now lie beyond the reach of our Philosophy. However, upon the Supposition of the three Spheres of Attraction and Repulsion just mentioned, nothing is more easy than to see how Solids may be converted into Fluids, and Fluids into Solids (as is done in Liquefaction and Freezing); for allowing that the first or innermost Sphere of Repulsion, is capable, like that of the Particles of Air, of being augmented by Heat, and diminished or totally suspended by Cold, it follows that Bodies must be more or less fluid in proportion to the degree in which they are affected by Heat or Cold. (Rowning, 1753, II, 5*n*–6*n*; italics in the original text)

3.2.3 An Attempt by Gowin Knight

According to Schofield (1970, 39), Rowning's conception, with its structure of "several concentric spheres of attraction and repulsion, surrounding the particles of bodies," was a peak in the dynamical corpuscularism. This high point was "soon to be inundated by the rising current of materialistic explanations remotely based on variations of Newton's aether."[19] However, as we have seen, a certain degree of materialism was already present in Hales and in Desaguliers—not with a reference to an aether as the carrier of the forces, but because of their tendency to let forces depend on attributes of matter. For Hales, air was somewhat special, and it alone was possessed with an "amphibious property". On the other hand, Desaguliers was far from generalizing his theory of the attractive and repulsive spheres of activity. In his treatment of elasticity as a sort of magnetic virtue, attraction and repulsion depend more on the structure of matter then being an "accident of distance".

Even Rowning's theory is not immune to this kind of materialism. As in Desaguliers' theory of the polarity-based elasticity, for Rowning, electricity and magnetism seem to elude the scheme of the alternating spheres of activities. According to him, electricity is another power (or active principle) "besides the General Powers [of Attraction and Repulsion] forementioned"—a power that characterizes only "some bodies". For example, "Amber, Jet [lignite], Sealing-Wax, Agate, Glass and most kinds of Precious Stones attract and repel light Bodies at considerable distances." In regard to magnetism, "the Loadstone is observ'd to have Properties peculiar to itself; as that by which it attracts and repels Iron, the Power it communicates to the Needle and several others." However, he did not go further in

[19]Schofield (1970, 15–16) distinguished between a first "mechanical" tendency of the dynamical explanations of matter and a second "materialistic" approach. They also reflect, partly at least, a chronological order, as the above-quoted passage indicates. More in particular, for the mechanists, "causation for all the phenomena of nature was ultimately to be sought in the primary particles of an undifferentiable matter, the various sizes and shapes of possible combinations of these particles, their motions, and the forces of attraction and repulsion between them which determine those motions. The materialists believed, instead, that the causes of phenomena inhere in unique substances, each possessing as an essential property the power to convey, in proportion to its quantity, some characteristic quality." Even if the distinction may be useful as such, I think that the situation was more fluid than in Schofield's account. My feeling is that *mechanism* and *materialism* were two extremes on a *continuum*, rather than two mutually exclusive options. For a slightly different criticism of Schofield's dichotomy, see Heimann and McGuire (1971, 234–235).

"feigning hypotheses" about their properties. As he explained in a footnote, "several solutions of the Properties of electricity and magnetism have been attempted by different Philosophers, but all of them so unsatisfactory as not to deserve a particular account in this Place" (Rowning 1753, I, 18–19).

A further reification of these somewhat materialistic tendencies would characterize the approach of Gowin Knight, some 15 years after Rowning began to publish his *Compendious System*. A fellow of the Royal Society since 1745, decorated 2 years later with the Copley Medal for his studies on magnetism, and the first Principal Librarian of the British Museum since 1756, Knight authored a small treatise bearing the remarkable, long title of *An Attempt to demonstrate, That all the Phænomena in Nature May be explained by Two simple active principles, Attraction and Repulsion: wherein the attractions of Cohesion, Gravity, and Magnetism are shewn to be one and the same; and the Phenomena of the latter are more particularly explained.*[20]

For Knight, the *Attempt* was the speculative outcome of an effort of "Simplicity of a Mathematical Demonstration" (Knight 1754, 2), rather than a quantitative study such as that of Hales or Desuguliers. Experimental results are reported sporadically, and mathematization mostly remains purely speculative.[21] The work seems to be conceived as an original, hypothetical-deductive system starting with three methodological propositions about knowledge and truth, an Axiom about God as the "first Cause of All things" and two definitions specifying what "immediate" and "mediate" causes are. From these follow other definitions and propositions (reaching 91 in number), often supplied with corollaries, which, in turn, refer to the previous statements. Attraction and repulsion are described as immediate causes, which are "acts of God himself" (*ibid.*, 5, Prop. X; 8, Prop. XV; 3, Def. I), and hence "the Effects of God's Will" (*ibid.*, 4, Coroll. II). As a direct emanation of God's Will, as Knight reasoned, every immediate cause "must necessarily be constant, immutable, and irresistible by any finite Force." This grants a peculiar status to attraction and repulsion: they ought to be constant, immutable, and irresistible as well. In Proposition XI, Knight offers an overview of *irresistible force*—which is, in fact, irresistible only at the point of contact:

> If two Particles indefinitely small, at a given Distance, attract each other with any given Force, how small soever; and that Force increases, as the Distances or the Squares of the Distances decrease, or in an higher Ratio: such Particles, in Point of Contact, will adhere with indefinite Force. *First*, Suppose the Force increases simply as the Distances of the Surfaces decrease; then at half the given Distance they will attract with double the Force, at half that Distance with four times the Force, at the next half with eight times, and so on . . .: in Point of Contact their attracting Force will have undergone an infinite Increase, and consequently be infinite. Now this will be the Increase, if the Force of Attraction is computed

[20]Knight (1754). Originally published in 1748, the book was then reprinted in 1754. I will quote from this edition.

[21]For example, Knight (1754, 44, Coroll. III to Prop. LXI) asserted that, even if we perceive gravitation as decreasing with the inverse square of the distance, it really diminishes "as the Distances increase," i.e., simply with the *inverse* of the distance, "on Account of having the Repulsive Principle compounded with it." However, as we shall see, the existence of two counter-acting attractive and repulsive principles is merely assumed.

> from the Surfaces; but since the Force is always to be computed from the Centers of the attracting Bodies, which in this Case are at an indefinitely small Distance from each other, the Force in Point of Contact, though not infinite, will be greater than any that can be assigned. And if this holds good in the simple reciprocal Ratio of the Distances, it is easy to perceive that it will much more do, if the Force be supposed to increase in a higher Ratio. (*ibid.*, 6, Prop. XI)

Of course, the same also holds for repulsion (*ibid.*, 9, Prop. XV). Now, having conceived attraction and repulsion as two contrary and equally irresistible causes that God, at the beginning, impressed upon each particle, they "cannot both, at the same Time, belong to the same individual Substance" (*ibid.*, 10, Prop. XIX). As a consequence, all matter is split into two qualitatively different classes, "one attracting, the other repelling" (*ibid.*, 10, Coroll. to Prop. XIX); note, though, that this is the only difference: "All the Primary Particles of Matter are originally of the same Size, and all round" (*ibid.*, 12, Prop. XXIV). Moreover, the repelling particles only repel each other, but they attractively interact with attractive matter. They "seem in respect to other matter also subject to the general Law of Attraction," thus adhering to the surfaces of the attractive particles.

With this premise, Knight developed a theory of matter based on a system of alternating attractive and repulsive spheres of *matter*. Let us imagine "the whole Infinity of Space ... filled with a Fluid, whose Particles were indefinitely small, and mutually repelled each other in the reciprocal Ratio of their Distances." Then, in Fig. 3.3,

> Let the Sphere *A* represent a Part of Space surrounded with such an elastick fluid, but itself quite void of repellent Particles; and let *B C D E* represent a Series of such Particles continued *ad infinitum*. The Particle at B being surrounded with repellent Particles on one Side, but having a Space void of them on the other, will be determined towards the Center *A*, and press upon the resisting Surface of the Sphere ... For the same Reason the Particle *C* will be determined towards the same Center, and press upon *B* till it has approached so near, that the repulsive Force of *B*, added to the repulsive Force of all the distant Particles beyond the empty Space about *F*, shall be equal to the repulsive Force of the infinite Series of Particles *D*, *E*, ecc. (*ibid.*, 16–17, Prop. XXXII)

Now, as Knight explained, "the repellent Particles of our elastic Fluid will be condensed round every corpuscle of attracting Matter, more or less, in Proportion to the Size of such Corpuscles" (*ibid.*, 18, Coroll. I to Prop. XXXIV). In other words, "every Corpuscles of attracting Matter will have round it ... as many repellent particles as will just balance its attracting Force" (*ibid.*, 19, Prop. XXXVI). Note that these will adhere as strongly as possible to the corpuscle (for the Prop. XI quoted above). Of course, this hinders the actual contact between corpuscles that can take place, and whenever Knight speaks of contact, "no more is meant than that they are brought as near as the Repulsion of their Surfaces will admit" (*ibid.*, Prop. XLV and Coroll. I, 25). Therefore, attractive corpuscles form compound bodies, and each of them is surrounded by adhering repulsive particles. This mix is called a primary corpuscle of a body by Knight, and "all the primary Corpuscles of Bodies are compounded of attracting and repelling Matter" (*ibid.*, 28, Prop. XLVIII). Cohesion is then explained in terms of a balance between attractive and repulsive matter, and a

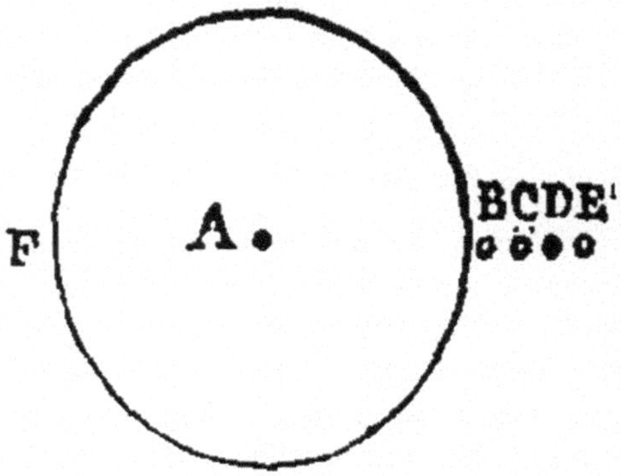

Fig. 3.3 An illustration of Knight's theory of matter as a system of alternating attractive and repulsive spheres. From Knight (1754, 17)

peculiar dynamical balance between the attraction and repulsion of the corpuscles constituting a body may also explain the cause of elasticity: "Attracting Corpuscles approach each other with an increasing Force, till they come so near, that the Repulsion at their Surfaces begins to counter-act their Force of Attraction ... and at a certain Distance the Repulsion at their Surfaces becomes so strong, as to equal and quite destroy the attracting Force" (*ibid.*, 35–37, Prop. XLIX and its Corollary). As a consequence, bodies are constituted more of voids ("pores") than of "solid parts" (*ibid.*, 37, Prop. L) and "the repellent Matter will be condensed partly in the Pores of solid Bodies, and partly upon their Superficies in Form of Atmosphere."

Knight also conceived of this structure as reproducing at every scale. Let us imagine that space were to be filled with a repulsive fluid; "if two or more Bodies be supposed to exist, and be placed at a Distance, their repellent Atmospheres will mutually put Bounds to each other on those Sides which mutually respect each other" (*ibid.*, Prop. LIX, 42). This should help account for the celestial motions and the structure of the solar system. The sun is surrounded by its repulsive atmosphere, which "is bounded by those of the Fixed Stars surrounding it." The balance of the attractive matter and the repulsive atmospheres surrounding each planet and star should preserve the Universe in a sort of dynamical equilibrium, hindering gravitational collapse and, with that, providing an answer to the cosmological paradox.[22]

[22]Knight's attempted solution to Newton-Bentley's cosmological paradox is included in the corollaries to Proposition XXX. God could "with equal Facility create the World infinite as finite" (Knight 1754, 15, Coroll. II to Prop. XXX). Let us assume that the world is infinite (quite an easy assumption for a Newtonian, as it was assumed by Newton himself in his correspondence with Bentley). In that case, "we have no need of a Cause to limit the Expansion of the repelling Matter, or to prevent the universal Conflux of the attracting. The first will disperse itself as equally as possible through every Part of Space: the latter may be disposed in such a Manner, that each System of Heavenly Bodies may be *in equilibrio* with those that surround it." On Knight's approach to the cosmological paradox, see Guicciardini (1996, 276–277).

3.2.4 The "Beautiful" Magnetic Theory of John Michell

Boscovich mostly ignored these theories, partly because he probably was not aware of them (as in Knight's case), and partly because of divergences on fundamental aspects. This was not entirely the case with the conception of another supporter of dynamical corpuscularism, with whom he became acquainted in 1760 when he visited England—namely, the natural philosopher (and English clergyman) John Michell.

In a long letter to his friend and pupil Giovan Stefano Conti (dated "Pera di Costantinopoli, 26 Febr. 1762"), Boscovich recalled that, a couple of years earlier, he had briefly discussed his own theory "only one evening in Cambridge . . ., since there was one man of Letters who told me to have believed for years that immediate contact of bodies does not occur, *but with none of the arguments I employ*."[23] One hint that could help identify the "man of Letters" is provided by Boscovich's journal during his stay in England, which includes a list of people whom he met in London, Oxford, and Cambridge. The last entry of the list reads "M. Michel, who does magnetic experiments," which most likely refers to John Michell (with *M.* simply standing in for the usual title of *Monsieur*).[24]

A letter to his brother Bartolomeo (dated November 20, 1760), which reports one of the meetings with Michell, confirms this occasion. Ruggiero particularly emphasized that his own "system has been discussed in various places, and I sketched it out on a slip of paper. Since a long time M. Michel is convinced, as he told me, without knowing anything of my theory, that immediate contact never occurs and is prevented by a repulsive force; but he does not prove this [i.e., *the repulsive force*] as I do."[25] In a previous letter, Boscovich recounted that Michell lent him his own book on magnetic experiments (most likely his *Treatise of Artificial Magnets*,

[23]"Solo una sera in Kembridge [*sic*], ed. era l'ultima della mia dimora in quella Università venne in occasione di darne [of his theory] una breve idea, essendovi un di que' Letterati, che diceva di aver creduto da varj anni, che non vi fosse l'immediato contatto de' corpi, *ma senza alcuno di quelli argomenti, che io adopro*" (Boscovich 2008, I, 62, italics added).

[24]*Ruggero Giuseppe Boscovich Papers*, 1711–1787, Bancroft Library, BANC MSS 72/238 cz. See, in particular, Series 5, Vol. 2: Notebook 1756–1766 (the files also include a register titled "*Gente conosciuta in Londra*", i.e., "People met in London," and other lists of people whom he met in Oxford, Cambridge and Tumbridge Wells). Together with the letters to Bartolomeo Boscovich, the journal helps reconstruct Ruggiero's stay in England: He arrived at Dover on the evening of May 23, 1760; on the 27th, he was in London after briefly visiting the Greenwich observatory. He left for Oxford on June 30 and later went to Cambridge. His scientific *grand tour* came to an end in late December, with a journal entry titled "Flessinga [Vlissingen, the Netherlands], 22 Dec: 1760." For an analysis of Boscovich's relationships with English scientists during his stay, see Proverbio (2003). On Boscovich and Michell, see also Heilbron (2015, 100–102).

[25]Ruggiero to Bartolomeo Boscovich, November 20, 1760: "Si è parlato anche un poco del mio sistema in vari luoghi, e ne ho lasciata una piccola idea in un mezzo foglio. M. Michel è gran tempo che è persuaso, come mi ha detto, senza aver inteso nulla della mia teoria, che non vi è mai l'immediato contatto, impedito da una forza ripulsiva; ma egli non la prova come fò io" (Boscovich 2010a, 402; note that "egli non *la* prova" certainly refers to the feminine *forza*).

originally printed in Cambridge in 1750 and in a second edition in 1751), and then explained his magnetic theory to him, which Boscovich described as "*bellissima*" (beautiful).[26]

In the original project of the *Treatise*, Michell had probably included a comprehensive theory of magnetism; however, as he put it in the introduction, "finding that this would swell these sheets to too great a bulk, I chose to defer that part till some other opportunity" (Michell 1751, 2). However, the opportunity never came, and the theory remained essentially unpublished. Nevertheless, some hints in the *Treatise* can help us reconstruct it *ex post*. To be sure, Michell rejected that magnetism could be reduced to a (double) fluid conception, and instead proposed that magnetic force acts at a distance according to Newton's inverse-square law (*ibid.*, 17–19). A crucial passage provides a more detailed account of his conception, which has sometimes been called a theory "of molecular magnets"[27]:

> If two Magnets be placed with their Poles of the same denomination together, they will damage each other considerably ... It is plain that, if we conceive any Magnet, as divided[?] into several, by Sections parallel to its Axis, that each of these will be endeavouring to damage all the rest. Now, if we suppose that the hardness of the Steel is able to resist this Endeavour, in some measure; this will very well account for any piece of Steel retaining its Magnetism to a certain degree, and for its not retaining any more than that: since after the power is become so great, as to be an overmatch for the resistance arising from the hardness of the Steel, the Magnet must necessarily reduce itself to such a power, as shall be just a balance for that resistance. And if we allow this reasoning to be just, the softer the Steel is, the less Magnetism it ought to retain, and the more easily it ought to retain. (Michell 1751, 12–13)

Another hint pointing to this as the core of Michell's theory emerged from Boscovich's above-quoted letter dated November 20, 1760. He reported that Michell conceived "iron as being formed by many tiny particles of lodestone, whereby an extreme attracts one point of any other similar particles according to the inverse square of the distances, and repels the other extreme ... In order to become a magnet,

---

[26]Ruggiero to Bartolomeo Boscovich, November 14, 1760: "The second day I had a lunch at the Trinity College, and in the evening I was at Dr. [Charles] Mason's home together with some friends; amongst other[s] there was Mr. Michel, a geometer and good physicist. The third day he [showed] me his beautiful magnetic experiments, and some other[s] which were new for me. Finally he very politely donated me 12 plates of steel—which he reduced to artificial magnets as I was there—as well as his printed book on this subject. He also told me his theory, which is beautiful." ["*Il secondo giorno pranzai nel Coll:[egi]o di Trin:[it]à, e fui la sera dal Dottor [Charles] Mason con varj amici, tra questi il Sig. Michel geometra, e fisico bravo, che il terzo giorno mi fece vedere le sue esperienze magnetiche assai belle, e varie giuntemi nuove, e al fine con estrema politezza mi regalò 12 lastre di acciaro, che aveva ridotte in calamita artificiale in mia presenza, e un suo libro stampato su questo dicendomi anche la sua teoria, che è bellissima.*"] (Boscovich 2010a, 402). This is the first meeting between Boscovich and Michell, probably on the evening of November 5, 1760: see McCormmach (2012, 66–67). The "reduction", i.e., Michell's method for preparing artificial magnets, is revealed in Michell (1751, 21 ff.). By the way, the Charles Mason mentioned was, according to the entries in Boscovich's "English" journal, the Woodwardian Professor of Fossils (Geology) at Cambridge.

[27]See Marković (1961, 130), Schofield (1970, 242), and McCormmach (2012, 67). As far as I can see, the expression *molecular magnets* does not occur in Michell's text and was probably coined by Marković (1961), to whom both Schofield and McCormach refer.

lots of those particles are turned to the same position, so that their actions do not destroy each other."[28]

3.3 Boscovich and the Newtonians

Whereas it is possible (and partly documentable) that Michell functioned as a bridgehead—of course, not the only one—for the penetration of Boscovich's ideas into Britain, primarily through his relations with Priestley (see Heilbron 2015, 102–103), Boscovich's consideration of dynamical corpuscularism remained poor. Desaguliers and Knight did not warrant any mention in Boscovich's works; yet, he met Knight in late July 1760 and attended some of his magnetic experiments, but, apparently, they did not discuss the theory of matter and Boscovich was most likely unaware of his *Attempt.*[29]

Boscovich probably read Buffon's French translation of *Vegetable Staticks*[30] and was acquainted with the theory of the fixed and elastic airs. This is evoked in Boscovich's (1749) dissertation *Sopra il turbine*, "on the whirlwind", as reported in the title, "which, in the night between the 11th and the 12th June 1749, greatly damaged Rome." The work is structured in three parts. After giving a detailed account of the events in the first part, the second part compares the Roman whirlwind with similar cases reported in the literature (especially in classical sources and in journey reports). Finally, the third part provides a taxonomy of the whirlwinds, mainly based on Aristotle's meteorology and Pliny the Elder, whereas their causes are investigated in the Halesian terms of rapid changes of air from an elastic, "volatile" state into a fixed state, and *vice versa* (see, in particular, Boscovich 1749, 172–175).

[28]"*Crede, che il ferro abbia moltissime particelle piccolissime di calamita, nelle quali un punto estremo attrae in ragion reciproca duplicata della distanza un de' punti di ciascun altra di simili particelle, e ripelle l'altro estremo, e il secondo suo punto estremo fa l'effetto contrario .... Il divenir calamita, crede, che non consista in altro, che in voltarsi di molte di queste particelle nella medesima posizione, sicché le loro azioni non si distruggono*" (Boscovich 2010a, 402).

[29]Knight was an important personality in mid-1760s London. Born in Corringham (Lincolnshire) in 1713, he was educated in Oxford. Elected a fellow of the Royal Society in 1745, he received the Copley Medal 2 years later for his study of magnetism; he was also a renowned compass maker. In 1756, he was appointed as Principal Librarian at the British Museum, an office that he held until his death in 1772 (for a sketched biography and his magnetic works, see Schofield 1970, 175–176). According to a letter to Bartolomeo dated July 21, 1760, Ruggiero Boscovich had met Knight the day before and described him as "famous, since he is the man who has invented, or improved very much, the art of making artificial magnets" (Boscovich 2010a, 336). He then attended Knight's experiments on July 24 (see Boscovich 2010a, 343, 347). In none of these letters did Boscovich mention Knight's *Attempt.* Moreover, from the correspondence emerges the fact that, before their meeting, Boscovich was entirely ignorant of Knight's name.

[30]S. Hales, *La statique des végétaux et l'analyse de l'air*..., ouvrage traduit de l'anglais par M. de Buffon, de l'Academie des sciences, Paris: Jacques Vincent, 1735. The Italian edition was published only 20 years later: *Statica dei vegetabili, ed analisi dell'aria*, Napoli: Raimondi, 1756.

However, Hales is introduced in a somewhat ambiguous manner:

> It is appropriate to mention something in advance, which the physicists have known long since, but that was clarified very much by the celebrated Stephen Hales in his *Vegetable Staticks*, a work poor in method [*opera scritta con scarso metodo*], but with lots of observations and incomparable reflections. If attentively read and meditated [upon], this offers a unique opportunity to know [*apre l'adito a conoscere*], with unequaled clarity, an exceptional number of the secrets of nature. (*ibid.*, 156)

The exact meaning of Boscovich's complaint about Hales' lack of method is unclear. However, his asserted partial discontent suggests that his consideration of the work might have been due to the theoretical aspects, rather than to the impressive amount of experimental data that it exposed. He also added that one should read Hales attentively and meditate on it so that the content of the work indirectly produces the knowledge of natural secrets. (Note that the Italian phrase "*aprire l'adito*", now highly unusual, literally means "to open the doorway" and "to make access possible.") This seems to imply that such knowledge is certainly stimulated by the content of Hales' book, but that it is also gained through personal reflection on the experiments and the way that Hales commented upon them, rather than being contained within the work itself.

In fact, in the subsequent paragraphs, Boscovich essentially describes his own theory, starting with Newton's conjecture that the particles of air are subject to a force of repulsion that is inversely proportional to the distances (i.e., the more two particles approach each other, the more they mutually repel each other.) He then emphasizes the concept of an inversion of the forces that he had explored as early as in the 1745 *De viribus vivis* (*ibid.*, 157–163). In other words, Hales' conception and his vocabulary become immersed in the context of Boscovich's mathematical theory of the forces. Elasticity is thus seen as an effect of the spatial distribution of the particles forming air, with the same mechanism that applies to all bodies:

> From this theory of mine can be seen very clearly something that can be gained from Newton's theory too, namely, in what manner a body can be transformed into another if the arrangement of its constitutive parts only changes. Let take as an example a fluid, elastic body like the air, each part of which endeavors to move away from its neighbor. The more the distances decrease, the more such repulsive force increases. However, if the distances are diminished as much as, at the smaller distance now gained, the repulsions are turned into attractions, those particles will not endeavor to move away anymore, but they will approach one another. (*ibid.*, 162)

Then, two modes in which elasticity can get lost are distinguished. In the first mode, "the particles of air, having approached one another to such small distances that repulsion is changed into attraction, that air would lose its elasticity and would transform into another body, or would acquire another form and other properties." In the second mode, if another substance nearby exerts a strong force of attraction on the particles of air, "there would be a compact part [*parte soda*] consisting of a particle of that substance and many particles of air, which would not lose their elastic force, but they have it constrained [*imprigionata*] in such a manner that . . ., as far as the effects are concerned, it would be as if [the elasticity] would get lost" (*ibid.*, 163). It is on this basis that Boscovich discussed Hales' experiments (mainly reported in *ibid.*, 166–171) and advanced a mechanical theory of the whirlwinds. They essentially consist of concentric layers of rarified (thus elastic) and fixed air in circular motion, which alternately attract and repel one another (*ibid.*, 171–175).[31]

To be sure, Hales' experiments fit well into Boscovich's scheme. It cannot be otherwise. For, in the latter's perspective, the conception of the elastic and fixed airs described in *Vegetable Staticks* was but a special case of his own general natural philosophy that was only based on universal attractive-repulsive actions among *every point* in nature. And, after all, an inspection of the passages from the *Theoria* where Hales is quoted confirms that the theory of the fixed and elastic airs was interpreted in consideration of Boscovich's theory of forces.[32]

Of course, one can view Boscovich's conjecture about a double mechanism governing elasticity, and thus causing the winds (*ibid.*, 171–172), as a much bolder hypothesis and as more demanding than Hales' speculation about the amphibious being of air. After all, Hales confined himself to describing experiments and to arguing about the proximal causes of the observed phenomena. Fixed and elastic air can even be considered as names for observables, which Hales instantiated with numerous experiments and observations about animals and vegetables. Speculation only concerned the idea that the elastic, repulsive air could compensate the attractive properties of the ordinary matter in order to preserve the dynamical equilibrium of the Universe. Boscovich, contrastingly, speculated about the unobserved mechanism of the observable elastic and fixed air.

Nevertheless, there is at least one important difference between Boscovich's conjecture—as demanding as it might be—and Hales' hypotheses, no matter how grounded in experiments and measurement we can consider them to be (not to mention the speculations of other Newtonians). I would call it a difference in *style*. Heimann and McGuire (1971) argued that, during the eighteenth century, Newton's concept of *active principles* went through a slow process of reification and was transformed into the idea of *active substances*. I am not entirely sure about the role Heimann and McGuire ascribe to Locke's *Essay Concerning Human Understanding* (1690) in this development, but I think that they are fundamentally correct in their description of active principles as features of the inherent activity of matter (or substances). This is clearly the case in regard to Knight, with his insistence on

---

[31]Of course, it is not the aim of this chapter, nor that of the book, to test the reliability of Boscovich's theory of the whirlwinds and confront it with eighteenth-century meteorology.

[32]Boscovich (1763, § 352): "Newton remarked that an air could from being volatile become fixed, & Hales especially gave a very full proof of this." § 379: "Hales demonstrated that many substances of the animal and vegetable kingdoms in a great part consist of air that has attained fixity." § 458: "Hales demonstrated by means of experiments that the great part of stones, which are produced in the bladder, and of the small ones in the kidneys, consists of pure air reduced to fixation; and that this can once again recover its volatile state. In this case, the compression of the air is not obtained simply by the boundaries that enclose it; for these would be completely broken down, since the air in such fixed solids is reduced to a volume that is even a thousand times less; and in this state, if the elastic forces still were unimpaired, all restraints would be easily overcome. Hales thought that, when in this state, it loses its elasticity; and this would indeed happen if its particles attained that distance from one another, in which there is no repulsive force, but rather an attractive force succeeds the repulsive force."

two kinds of matter: one attractive and one repulsive (and let us remember that he describes attraction and repulsion as two "simple active principles", even in the title of the *Attempt*). However, the reification was already operating in Hales' idea of amphibious air and in his implicit distinction between two qualitatively different kinds of matter: elastic and inelastic. From this point of view, Hales and Knight embody two successive phases of the reification process that has increasingly turned Newton's active principles into the active substances of, say, a late eighteenth-century natural philosopher like Priestley.

Michell's case adds further evidence, and a comparison with Boscovich may shed light on their respective peculiarities. Both maintained that matter, up to its elementary components, is not subject to immediate contact, for this "is prevented by a repulsive force." But Boscovich introduced the repulsive force as early as in the 1745 *De viribus vivis* based on a general principle of continuity that should be preserved—something that, in the *Theoria*, he would call an undisputed and unavoidable metaphysical assumption. Michell, instead, started from assumptions about the constitution of matter that Boscovich would have considered as arbitrary hypotheses regarding an unexperienced microscopic nature of bodies (not to speak of Knight's reification of attraction and repulsion into two radically different kinds of matter).[33]

Let us remember that, in Boscovich's report, Michell conceived iron as actually consisting of magnetic particles formed by qualitatively different poles: given two particles, A and B, of iron or lodestone, both endowed with poles *n* and *s*, A*n* attracts (say) B*s* and repels B*n* and A*s* attracts B*n* and repels B*s*, always according to the inverse square of the distances. For Michell, magnetism is the combined effect of the distances and of a quality inherent to each pole, for A*n* can never attract B*n* or be repelled by B*s*. In the *Theoria*, instead, Boscovich gave a different, merely *positional* interpretation of polarity:

> With regard to attraction, it is clear that this can be present in the particles, and that it must depend upon their combination [*ab earum textu*]. Moreover, there are very many phenomena

[33]This attitude also emerges in the dedicatory epistle of the *Theoria*: "I put on one side all prejudice, and started from principles that are undisputed and indeed commonly accepted; I used perfectly sound arguments, and by an uninterrupted chain of conclusions, I arrived at a single, simple, continuous law for the forces that exist in Nature. Its application showed me the constitution of the elements of matter, the laws of Mechanics, the general properties of matter itself, and the chief characteristics of bodies, in such a manner that the same uniform method of action in all things disclosed itself at all points; being deduced, not from arbitrary hypotheses, and fictitious explanations, but from an uninterrupted chain of reasoning only [*Omni praejudicio seposito, a principiis exorsus inconcussis, et vero etiam receptis communiter, legitima ratiocinatione usus, et continue conclusionum nexu deveni ad legem virium in Natura existentium unicam, simplicem, continuam, quae mihi et constitutionem elementorum materiae, et Mechanicae leges, et generales materiae ipsius proprietates, et praecipua corporum discrimina, sua applicatione ita exhibuit, ut eadem in iis omnibus ubique se prodat uniformis agendi ratio, non ex arbitrariis hypothesibus, et fictitiis commentationibus, sed ex sola continua ratiocination deducta*]" (Boscovich 1763, 8–9). In view of this, I cannot share Thackray's (1970, 145) assertion that (especially with reference to Knight) Boscovich "a decade later . . . was to pursue the issue to its logical conclusion."

of magnetism, which will show that magnetic force is generated by changing the disposition of the particles, or is destroyed, or more frequently is augmented or abated; examples of this everywhere come under the observation of those who study magnets. Also, poles that are attractive on one side and repulsive on the other, which occur in magnetism, agree with my theory; for, *the sum of the forces on one side may be greater than the sum of the forces on the other*. A somewhat greater difficulty arises from the huge distance to which this kind of force extends. But even this can take place through some intermediate kind of exhalation, which, owing to its extreme tenuity, has hitherto escaped the notice of observers, and such as by means of intermediate forces of its own also connects remote masses; *if perchance this phenomenon cannot be derived from merely a different combination of points having forces represented by my curve itself*. (Boscovich 1763, § 515; italics added)

To be sure, the microscopic structure of matter also plays a role in Boscovich's natural philosophy, but only as far as the positional properties of matter and its components are concerned. Poles are not qualitatively different; they are merely points combined in a special manner, which allows for repulsion on one side of each particle and attraction on the other side, according to Boscovich's curve. Moreover, notwithstanding its asserted beauty, Michell's theory was not even mentioned in the 1763 edition of the *Theoria* (also note that the above-quoted passage from § 515 in the 1763 edition, after his meeting with him, is identical to that of the first version: see Boscovich 1758, § DX).

Whereas Michell demanded ontological commitment, Boscovich was committed to the physico-mathematical style and its implied mathematical neutralism. Let us instantiate this with Boscovich's complex relationship with what he calls "powers". Heimann and McGuire (1971, 235–236) pointed out that, in the early Newtonian tradition, the concept of powers occasionally became associated with that of active principles (see, e.g., Rowning 1753, II, 12). According to them, this represents "a conception of nature which underlies both the tradition of imponderable fluids and that of interparticulate force." In this conception, "to ascribe a power to a material object is to assert what it can or cannot do in virtue of its intrinsic nature . . . Powers were conceived as being substantively present in entities, thus defining the entities' essence in terms of inherent activity" (Heimann and McGuire 1971, 235–236). As we have seen in the previous chapter, the notion of powers occurs in Boscovich in the Latin form of *potentiae*, obviously mediated by Newton's *Opticks* ("*Annon exiguae corporum particulae certas habent virtutes, potentias, sive vires*" reads the Latin formulation of the last Query). Powers could be considered the causes of the actions or forces (see Boscovich 1745, § 13; Boscovich 1755b, 300); they can be inherent to the nature of the bodies or added to and separable from them, or they may depend on the Creator's free will. However, we can remain agnostic about this and take advantage of the neutral feature of mathematics in order to harness the supposed powers into a unique mathematical structure.

On the one hand, there was no reason to be overly harsh toward the Newtonians. In the *Theoria*, they are identified with the "modern philosophers who seem to admit impenetrability and active forces . . . as the primary properties of matter placed in its very essence [*activas vires. . . pro primariis materiae proprietatibus in ipsa ejus essentia sitis*]"; however, the law of forces can "be used in all these kinds of philosophising, and can be adapted [*aptari*] to the mode of thought peculiar to any one of them" (Boscovich 1763, § 516). On the other hand, *adapted* does not mean *identical*. Contrary to the Newtonian theories of matter revealed in the previous section, Boscovich's conjectures are not guesses about powers and their carriers; rather, they are guided by the same epistemological position—ontological agnosticism associated with mathematical neutralism—that proved to be crucial for his theory of forces.

3.4 Vexed Questions

What I have discussed in the previous sections is not the only sense in which phrases like "dynamical theory of matter", "dynamicism", and related expression have been and are currently employed by many historians of scientific and philosophical ideas. Another important meaning is connected to Leibniz's conception of force, especially with his *Specimen dynamicum* (1695). This is particularly relevant to the present discussion, because it has often been maintained that Boscovich's natural philosophy, because of the relevance and the primacy of the notion of force over other concepts, is, in truth, merely a variation of a Leibnizian dynamistic "force". In turn, this association of Boscovich with Leibniz has often been viewed as a fundamental step in the (pre)history of the field theory. In the following subsections, I will consider dynamicism in a broader sense and shall discuss these two closely related and highly popular historiographic theses, namely, the influence of Leibniz's concept of force on Boscovich and the influence of Boscovich's concept of force on Faraday.

3.4.1 Leibnizianism Disguised?

In the years that followed the publication of Child's 1922 edition of the *Theoria*, an increasing number of scholars became interested in Boscovich's main work in natural philosophy. Partly reacting to the phenomenalistic perspective expressed in the editor's introduction, many of them pointed out that, in Boscovich's theory, the features of matter—first of all, impenetrability—are explained in terms of an interplay of forces, whatever the term "force" may denote. An influential representative of this tendency is Max Jammer, in his classical *Concepts of Force* (Jammer 1957, 177–178). He recognized that "strictly speaking, the ordinates represent accelerations only . . . From the standpoint of physics, however, we may claim that a physical theory, based on the notion of force as its most fundamental conception, may be called 'dynamic', even if it does not interpret its fundamental concept as a metaphysical entity."

According to Jammer, this is sufficient to proclaim Boscovich's theory as the best example of a dynamic conception of matter in Leibniz's spirit (because of the insistence on material points as "force-centers"). And, after all, does not Boscovich

(1763, § 1) describe his own "system" as a "midway between that of Leibniz and that of Newton"?

From this perspective, it might well be that Boscovich implicitly started from some strong assumptions regarding forces "existing in nature", as the title of the work seems to insinuate. In addition, in all of his works of natural philosophy, Boscovich made numerous references to Newton's *Opticks*, particularly Query 31, wherein hypotheses about the nature of forces are expressed. Many scholars began to suspect that Boscovich's cautious "Newtonian-style" warnings about the reality of forces simply obscured his most profound insights. Influential writings expounding this "force-friendly" view cover a noticeable span of years and include works by Želiko Marković (1961), Lancelot Law Whyte (1961), Robert Schofield (1970), Arnold Thackray (1970), George Gale (1974), John Heilbron (1979 and 2015), Mirko Grmek (1996), Edoardo Proverbio (2003), Michael Friedman (2004 and 2013), and Russell McCormmach (2012). To be sure, a complete list would be far longer and would include many of the contributors to the proceedings of past Boscovich conferences, as well as such far-reaching studies as Mary Hesse's (1961) *Forces and Fields*. Many of them also insist that Boscovich's theory fits well into the context of so-called dynamicism or, more precisely, dynamical theories of matter broadly understood—something that includes, but is not limited to, the dynamic-corpuscular conceptions that I have considered so far.

Indeed, it is not easy to characterize dynamicism, and extant definitions are unsatisfying to say the least. Jammer (1957, esp. Chap. 9) placed Boscovich alongside Leibniz, Kant, and Spencer under the common label of "dynamism". He quoted a phrase from Leibniz's *Specimen dynamicum* as its motto: "*Agere est character substantiarum*" ["It is the character of substance to act"]. According to Jammer, Leibniz's conception can be portrayed using another maxim: "*Quod non agit, non existit*" ["What does not act, does not exist"], "in contrast to the traditional '*Operari sequitur esse'* ['Action follows existence']."[34] Both sentences could also be applied to Boscovich, to whom Jammer (1957, 170) ascribed "the real Leibnizian theory of dynamics." This is true, as, even if his concept of force was "merely relational or functional ... Since impenetrability and extension, in Boscovich's

[34]Jammer (1957, 169–170). "*Agere est character substantiarum*" is quoted from Leibniz (1695, 235/435). "*Quod non agit, non existit*" is traditionally ascribed to Leibniz but is possibly extrapolated from a sentence included in a manuscript allegedly from 1673–75, *De vera methodo philosophiae et theologiae ac de natura corporis*. While treating the corporeal nature in general (which, according to him, cannot be reduced to the Cartesian *res extensa*) Leibniz stated: "*Corpus ergo est Agens extensum: dici poterit, esse substantiam extensam, modo teneatur omnem substantiam agere, et omne agens substantiam appellari. Satis autem ex interioribus metaphysicae principiis ostendi potest,* quod non agit, nec existere, *nam potentia agendi sine ullo actus initio nulla est*" ("A body is therefore an extended agent. It can be said that it is an extended substance, only if it be held that all substance acts, and all agents are substances. It can be shown adequately from the essential principles of metaphysics that *what does not act does not exist*, for there is no power of acting without a beginning of action." Quoted from Leibniz 1673, 158, emphasis added; translation in Leibniz GW, Philosophical Papers and Letters. Springer, Dordrecht, 1969, p. 271, n. 9). I am indebted to Stefano Di Bella for a discussion on this Leibnizian passage.

view, are merely spatial expressions of forces, 'force' is consequently more fundamental than 'matter' ... Boscovich's theory of physical sciences may therefore be rightly called 'dynamic'" (Jammer 1957, 178).

The statement that "force is more fundamental than matter" is probably too wide and generic for the purpose of labeling any conception as a variation on the Leibnizian dynamicism-theme. After all, it depends on the notion of "fundamental". If this is taken in the strongest sense, i.e., that forces are the "fundament" of matter, it would be an apt description of Leibniz's doctrine. As for Boscovich, he would reply that the primary elements of matter are indivisible, homogeneous, unextended material points *endowed with* forces.[35] But, of course, this claim does not entail that forces are the substance(s) of matter. The question as to whether forces are superadded by a Creator or are instead intrinsical to matter is not answered here—in fact, it is not even posited. As we saw, Boscovich always expressed an agnostic attitude toward this question. Therefore, one is forced to consider a weaker sense of the statement "force is more fundamental than matter," which only entails the primacy of force over matter. But this would also be applicable to some tenets of Newton's *Opticks*[36] and would potentially be in conflict with Leibniz's *Specimen dynamicum*, whose tenets are much more demanding.

More generally, the concept of activity as an inherent quality of substances is one of the distinguishing features of the Leibnizian view.[37] Now, it is not my goal to discuss or compare such concepts in the Newtonian vs. Leibnizian context. The main difference between the two approaches can be briefly sketched out as follows: whereas, in the Newtonian tradition, matter is what passively resists motion, thus it is not conceived as an inherently active substance (hence "active principles", whatever they are, should be added by God in order to facilitate motion, as well as fermentation, generation, and the phenomena described in the last Query to the *Opticks*), this was a common tenet within the Leibnizian tradition, and also applied

[35]See Boscovich (1763, §§ 7–9): "The primary elements of matter are, in my opinion, perfectly indivisible and non-extended points .... As an attribute of these points, I admit an inherent propensity to remain in the same state of rest or of uniform motion in a straight line .... I therefore consider that any two points of matter are subject to a determination to approach one another at some distances, & in an equal degree recede from one another at other distances. This determination I call *force*; in the first case, *attractive*, in the second case, *repulsive*." Homogeneity is outlined later in the *Theoria* (Boscovich 1763, §§ 92–99). See also Boscovich (1757, § 90), in which he explicitly identified his "*puncta*" with Newton's "*primae particulae*" and emphasized that they are unchangeable.

[36]On the primacy of force in some periods of Newton's thought, see McGuire (1996, 193–194) and Ducheyne (2014, 684–685, 690–692).

[37]Leibniz's crucial contributions in this respect are the *Specimen dynamicum* (Leibniz 1695) and the texts published in Costabel (1960); see also Leibniz (1994). Activity and force as main tenets of the whole dynamics of Leibniz are discussed at length in Gueroult (1934), Costabel (1960, esp. 6–14), Iltis (1970b), Gale (1970, 1973), Most (1984), Garber (1985), Hacking (1985), Gale (1988), Duchesneau (1994), Garber (2009, esp. 99–179), Fichant (2016), and Tho (2017a, b).

to some critics of Newtonianism in Britain, whose possible connection to Leibniz is somewhat obscure.[38]

The articulation of this difference is far from sufficient to account for every aspect of Newton's and Leibniz's physics, and can be detailed more and more—which, again, falls outside of the scope of this book. But even in this rough form, it signals that there are at least two very different senses of activity in the two traditions that allegedly are most relevant to Boscovich: one—the Newtonian—in which active principles are superadded to matter, and the other—the Leibnizian—in which matter is intrinsically active. Hence, a vague and ambiguous reference to something "active"—as in Jammer's reconstruction—is not sufficient to view Boscovich as a more or less radical dynamicist in the Leibnizian tradition. On the other hand, if activity is taken in the most proper Leibnizian sense as the action of a force that "constitutes the inmost nature of the body" (Leibniz 1695, 235, 435), hence grounding the idea of an inherently active substance, this is completely foreign to Boscovich.

Perhaps there is no better source than Leibniz's *Discourse on Metaphysics* for us to appreciate the difference with Boscovich's non-dynamicist (in the sense of Leibniz) conception:

> But the force or proximate cause of these changes is something more real [than the changes of motion that it induces], and there is sufficient basis to attribute it to one body more than to another. Also, it is only in this way that we can know to which body the motion belongs. Now, this force is something different from size, shape, and motion, and one can therefore judge that not everything conceived in body consists solely in extension and in its modifications, as our modems have persuaded themselves. Thus we are once again obliged to reestablish some beings or forms they have banished. And it becomes more and more apparent that, although all the particular phenomena of nature can be explained mathematically or mechanically by those who understand them, nevertheless the general principles of corporeal nature and of mechanics itself are more metaphysical than geometrical, and belong to some indivisible forms or natures as the causes of appearances, rather than to corporeal mass or extension. (Leibniz 1686, 444/51–52)

Boscovich would perhaps have agreed that "not everything conceived in body consists solely in extension and in its modifications"—but for reasons connected with his Newtonianism and the more or less implied commitment to the active principles of *Opticks*. He would most likely also have agreed that "the general principles of corporeal nature and of mechanics itself are more metaphysical than

[38]The bibliography on the concept and role of active principles in Newton is substantial and, at least in some cases, discordant. Scholarly work converges in maintaining that (quoting Ducheyne 2014, 691), "according to Newton, bodies are intrinsically passive and are moved only by active principles, that is, non-mechanical agents." (Note that Ducheyne, *ibid.*, 690–695, also gives a detailed account of Newton's idea of active principles.) But, once it is recognized that active principles for Newton are (or refer to) causal agents of forces, there is no consensus on the fine structure of the causation (see Joy 2008, 93–104). On active principles in Newton, see, Westfall (1971, 310–319, 363–400), Heimann and McGuire (1971), McMullin (1978, 43–56), Heimann (1978), McGuire (1996), McMullin (2001), and Ducheyne (2014, 690–695). On active principles after Newton, see Thackray (1970) and Heilbron (1979, 46–71).

geometrical": a clear-cut distinction between metaphysics and mathematics was an obvious tenet for a Jesuit, after all. However, with his notion of force as mutual determination, Boscovich would hardly have accepted that forces, as the "proximate causes" of the changes in motion, are in any sense "more real" than the motions themselves. Moreover, with force being a mutual action, he would not see any "sufficient basis" in order "to attribute it to one body more than to another" in Leibniz's sense (for a comment on this, see Garber 2009, 154–155).

I have used this long quotation from Leibniz's *Discourse on Metaphysics* because this is the passage that, according to Gale (1988, 64), "shows a plausible description of the processes leading Leibniz to his discovery of the interpretation of 'force' as *vis viva*." Thus, in Gale's account, it is crucial in the development of Leibniz's dynamical program. However, if he is right, the displayed quotation proves that the context of Leibniz's and Boscovich's understanding of force is completely different (*contra* Gale 1974). How should we understand, in Boscovich's terms, Leibniz's reference to "force or proximate cause"? If we use Boscovich's notion of force, Leibniz's statement is as far as possible from Boscovich's view. If forces are determinations, how could they possibly be "more real" than the change of motion that they induce? On the other hand, if we translate Leibniz's force with Boscovich's power, then powers are entities (whose real nature is entirely unknown) in their own right. Therefore, we cannot simply "attribute [a power] to one body more than to another."

3.4.2 A Prototheory of Field?

There is still another strategy for associating Leibniz and Boscovich—not through the concept of force, but through the Leibnizian monadology. Gale (1974, 41–48) emphasized that Boscovich was aware of the influence that Leibniz's idea of the monads exerted. To epitomize this, he quoted the following passage from the "Synopsis of the whole work" (Boscovich 1763, 23): "I am not the first to introduce indivisible & non-extended points into physical science; for the 'monads' of Leibniz practically come to *the same thing*" (emphasis added). However, unfortunately for Gale, in this passage "the same thing", i.e., the same result that is reached by both Leibniz's monads and Boscovich's material points, is only the critical idea that the "first elements" of (extended, according to Boscovich) matter have no extension. After this qualification, any similarity between points and monads promptly disappears, as Boscovich recognized since his (written in 1748, but not published until 1757) *De materiae divisibilitate* (Boscovich 1757, § 12 and footnote *e*; see also Boscovich 1763, §§ 138–139).[39]

[39] Of course, there are more structural reasons for why Boscovich's points are *not* Leibniz's monads. In particular, here, Gale missed two important points: first, that Boscovich might have developed his own idea of points from other sources, probably as a reaction to "atomistic" ideas that had experienced some fortune among the Jesuits, and this led him to be suspect of Leibnizian monadology (I will return to this in the next chapter); second, the crucial importance of the

In the same article, Gale particularly insisted on the Leibniz-Boscovich-Faraday connection, trying "to argue for the claim that Leibnizian conceptions are significant primary ancestors in the evolving chain of concepts which lead up to the theoretical systems we now call 'field dynamics'" (Gale 1974, 28); of course, he considered Boscovich's natural philosophy as one of such conceptions. I am not sure that a chain of this kind actually exists (and I do not intend to take a position on this), but I would certainly challenge the belief that Boscovich belongs to it. Contrary to Gale (1974, 41), I cannot take for granted that "Boscovich is justly recognized as one of the main figures involved in the founding of field theory" and that there is "no good reason, both structural and historical, to dispute this interpretation."[40] In the following paragraphs, I will try to offer some "good reasons" to dispute it.

The most significant advocate of Boscovich's natural philosophy as a premise to Faraday's field-conception was Faraday himself (e.g., Faraday 1844), whose Boscovichean tendencies have been noted by many commentators old and new (for some early examples, see Heimann 1971, 235 n. 2). It should also be remembered that, while Boscovich's theory had not received much attention from his scientific colleagues in continental Europe (Baldini 2006, 406), in Britain, it soon raised discussions. In fact, it greatly influenced the scientific context in somewhat unexpected ways through figures such as Priestley and Davy. Due to their mediation, it finally reached Faraday (see, for example, Olson 1969, Feingold 1993, Harmon 1993, Heilbron 2015). In his biography of Faraday, L. Pearce Williams (1965, esp. 53–94) particularly insisted on this background, then claiming that much of his field-theoretic conception would be a development of Boscovich's "point atomism" (Williams 1965, 127–128).

Faraday's regard of Boscovich as a precursor is a matter of fact (e.g., Faraday 1844, 289–291). But to assert that this is a historical connection *per se*, independent of Faraday's self-presentation, is an entirely different issue, which should be historically and structurally proven. In fact, Spencer (1967) and Heimann (1971) have shown that Faraday's interest in Boscovich emerged later as a justificatory attempt to

characterization of forces as *determinationes*, which prevents his points from being considered as *immaterial* force-centers. After all, Boscovich referred to them as *materiae puncta*, points of *matter*. This is somewhat opposed to Leibniz's "dynamical viewpoint", as later explored by Gale (1988).

[40]Of course, I can agree, at least partly, with more cautious readings—i.e., as far as Boscovich is not viewed as "one of the main figures involved in the founding of field theory," but rather as an actor in a more complex, not necessarily linear, historical-theoretical process. Michael Friedman has advanced a broader description of dynamicism as characterizing conceptions in which "the basic properties of solidity and impenetrability are not taken as primitive and self-explanatory, but are rather viewed as derived from an interplay of forces" (Friedman 2004, x). This would let us associate, among others, Kant's *Physical Monadology* and *Metaphysical Foundations of Natural Science* with Boscovich's *Theoria*, and might suggest viewing the theory of matter of the former, at least, "as an important step in a gradual process of transformation from the more 'passive' and 'mechanical' conception of matter prevalent in the seventeenth century to an 'active' and 'dynamical' conception characteristic of the nineteenth century, when the concepts of energy, force, and field finally triumph over those of (primitive) solidity, (primitive) impenetrability, and (absolute) indivisibility" (Friedman 2013, 97).

defend his own theory. They have also contended that some aspects of Faraday's theory are inconsistent with Boscovich's views on the structure of matter. For example, Faraday seemed to distinguish between matter and the material points, which is an idea that Boscovich would not accept (Spencer 1967, 199; Heimann 1971, 244–245). Moreover, Faraday (1844, 291) strongly claimed that matter was continuous ("Matter will be continuous throughout, and in considering a mass of it we have not to suppose a distinction between its atoms and any intervening space." See also Heimann 1971, 245–246)—again, something that Boscovich explicitly refused. There is enough to argue that Faraday misinterpreted Boscovich's theory on fundamental aspects, including the idea of force as a mathematical determination, which, of course, for Faraday, is a physical entity (something that Boscovich would have instead called a power).

More precisely, Faraday (1844, 290) maintained that "his [i.e., Boscovich's] atoms, if I understand aright, are mere centres of forces or powers, not particles of matter, in which the powers themselves reside." Of course, in Boscovich's view, powers do not reside within the material points (in fact, they elicit forces, i.e., external actions), whereas Faraday makes no distinction between powers and forces. Moreover, for Boscovich, points *were* the elementary *material* components of matter, to the extent that there is virtually no difference between "the mass of a body" and "the total quantity of matter pertaining to that body" (see Boscovich 1763, § 378).

Here, in addition to the evidence from the texts emphasized by Spencer and Heimann and to Boscovich's own concept of a material point as I described it in Sect. 3 of the present chapter, I suggest that the inspection of the curve of forces highlights another important difference, which makes it structurally incompatible with Faraday's conception, as well as with other field-theoretic versions. Boscovich's view, like Newton's, only makes sense if there are at least two points with force being exerted on the line of their junction—in other words, if there is a distance between points. If one point is annihilated (if there is no distance to be measured), then force is annihilated as well. Contrastingly, if we conceive points as force-centers, then force should be "monopolar" and persist even in the case of one point (which is actually a center of force) being annihilated. Note that this was Faraday's original understanding of force:

> The notion of the gravitating force is, with those who admit Newton's law, but go with him no further, that matter attracts matter with a strength which is inversely as the square of the distance. Consider, then, a mass of matter (or a particle), for which present purpose the sun will serve, and consider a globe like one of the planets, as our earth, either created or taken from distant space and placed near the sun as our earth is;—the attraction of gravity is then exerted, and we say that the sun attracts the earth, and, also, that the earth attracts the sun. But if the sun attracts the earth, that force of attraction must either arise *because* of the presence of the earth near the sun; or it must have *pre-existed* in the sun when the earth was not there. If we consider the first case, I think it will be exceedingly difficult to conceive that the sudden presence of an earth, 95 millions of miles from the sun, and having no previous physical connexion with it, nor any physical connexion caused by the mere circumstance of juxtaposition, should be able to raise up in the sun a power having no previous existence. As respects gravity, the earth must be considered as inert, previously, as the sun; and can have

> no more inducing or affecting power over the sun than the sun over it: both are assumed to be *without* power in the beginning of the case;—how then can that power arise by their mere approximation or co-existence? That a body without force should raise up force in a body at a distance from it, is too hard to imagine. (Faraday 1855, 571–572)

He contrasted this view, for reasons also involving—according to him—the "conservation of force", with the one that he endorsed, "namely that the power [meaning the "gravitating force"] is always existing around the sun and through infinite space, whether secondary bodies be there to be acted upon by gravitation or not." Far from being restricted to gravity, this is a general model. It is, in fact, "in philosophical respects, the same as that admitted by all in regard to light, heat, and radiant phenomena and (in a sense even more general and extensive) is that now driven upon our attention in an especially forcible and instructive manner, by the phenomena of electricity and magnetism" (*ibid.*, 574).

Faraday's parallel with gravitation was reprised by Maxwell (1865, esp. 492–493), who tried a more technical approach and developed it further, remarking that "the lines of gravitating force near two dense bodies are exactly of the same form as the lines of magnetic force near two poles of the same name; but whereas the poles are repelled, the bodies are attracted"; he could even obtain equations for what he called "the field of gravitation".[41] Two decades later or so, Heinrich Hertz, in a series of lectures that he delivered at the University of Kiel in 1884, came back to the gravitational analogon as an illustrative example of the conceptual foundations of the theory of field. His words may serve here as a clarification of an important common aspect of Faraday's and Maxwell's visions within this framework:

> Let us take the attraction, as itself, that the Sun exerts on the planets revolving around it. According to the pure theory of immediate distant action [i.e., Newton's theory] that force does not act through void space ... Instead, it is an external force exerted on the planets. It only acts in the places at which a planet is found ... Moreover, it is acting only until planets are present; if we remove them completely, the action of the sun does not take place anymore, the whole space surrounding it is void and indifferent and the gravitational effect of the Sun is only confined to itself. Therefore, according to this conception, in order that we can have a gravitational effect [*damit überhaupt von einer Gravitationswirkung die Rede sein könne*], two bodies are necessarily required, none of which is, in principle, the attracting or the attracted body, both being instead coordinated [*coordinirt*]. If we remove one of them, not only is the effect not perceptible—it is absolutely not present, it is absolutely not conceivable. On the contrary, according to the theory that the force is mediated, things go as follows: planets are attracted toward the Sun because they are within a space that pulls them against it because of a peculiar state in which the space is placed ... The space is put in such state by the Sun, not by means of a distant action but through action from particle to particle. More precisely, the sun does not put only the space occupied by planets in such a state, but the whole space, and the gravitational effect of the sun would not be different from how it actually is, not only in the case that a planet be present but also if none of them should be there. "Attracting body" and "attracted body" have not the same meaning: a body can

[41] As known, Maxwell's model of the gravitational field was extremely poor and unsatisfying. On Faraday's and other "anomalous views of gravitation" in this context, see also van Lunteren (1991, esp. 147–179).

attract without being attracted, and *vice versa*. Even if a body can include both features, nevertheless the two effects can conceptually be distinguished. (Hertz 1999, 61–63)

Let us come back to Boscovich and compare his view with that of Faraday and with Hertz's interpretation of the field-theoretic conception. If my picture of Boscovich's idea of force as determination is correct, then—contrary to Faraday's expectations—he must not have found it *exceedingly difficult to conceive that the sudden presence of an earth ... should be able to raise up in the sun a power having no previous existence*. In Boscovich's view, the earth *must be considered as inert, previously, as the sun*; to Faraday's concerns, expressed in his question: *How then can that power arise by their mere approximation or co-existence?*, he could easily counterpose his agnosticism as regards powers. Moreover, for Boscovich, *attracting body* and *attracted body* do certainly have the same meaning—or, in Faraday's own words, for him, the earth *can have no more inducing or affecting power over the sun than the sun over it*. Finally, in Boscovich's conception of the force, space ought simply not to be put in any state whatsoever, for *real* space, in Boscovich's understanding, is nothing but the indifferent scenario of coordinates where physical effects actually take place—and, to make things worse, real space was, for him, as discrete as matter.

Of course, we are free to interpret Boscovich's curve of forces in light of the notion of potential, this being a function that associates every point of space with the value of the force acting on a mass at that point. This will probably offer an accurate description of Boscovich's theory, as it does when we use potentials to describe how Newton's gravitational force changes throughout a given space. This is a piece of standard physics that can be learnt virtually from any handbook—but it is neither philosophy nor history of physics. Rather, it is part of a well-known mechanism by which physicists neglect theoretical changes or, in Kuhnian terms, by which normal science is erected, day by day, through the creation of heroes and precursors.

Summarizing, Faraday greatly underestimated or completely overlooked an impressive number of crucial aspects of Boscovich's conception. Instead, he tended to read it in light of a theory that he built independently, before he knew of Boscovich. One reason certainly was his lack of direct knowledge of the original texts, but there are still other causes (Heimann 1971, 237; Giusti Doran 1975, 165–170). In any event, there are fundamental aspects that make the two theoretical frameworks—namely, Faraday's and Boscovich's—incompatible. In this sense, Faraday operated within a sort of distortion of Boscovich's natural philosophy. Such a distortion, however, is the *explicandum* and not the *explicans*: it cannot be invoked as evidence to argue for Boscovich's influence on the development of the field-theoretic view, but should be explained in view of, e.g., the British context and the peculiar reception of Boscovich's theory in Britain (and elsewhere).

Chapter 4
The Book of Genesis

4.1 A Research Program from 1748: The Camaldolese Ur-Theorie

Having explained why I do not see Boscovich's conception as belonging to the corpuscular-dynamical theories of matter that flourished in eighteenth-century Britain, and why he can hardly be associated with "dynamicists" of various kinds, I will now proceed to explore *his* theory of matter as it emerges from his writings and in its proper context.

It was probably around 1748 that Boscovich began to give serious consideration to the theory of matter. This new interest clearly emerged in the second part of the dissertation *De lumine*, issued in September 1748. The first part provided a more or less standard presentation of geometric and Newtonian optics, which Boscovich wrote so that a pupil—Andrea Archetti—could defend it at a public occasion.[1] However, Boscovich himself debated the second and arguably more delicate part on September 5. Here, he reprised and attempted to develop further the theory of repulsive and attractive forces that he had revealed 3 years earlier.

At the very beginning of *De Lumine*, Part II, the curve of forces is mentioned to explain "the mechanical causes" of the properties of light (particularly the rectilinear propagation, refraction, reflection and diffraction). However, after commenting on the graph of the curve, Boscovich also suggests that the properties of matter themselves follow from its "form", i.e., the constellation of geometric properties of the curve:

[1]This is the complete title as taken from the frontispiece: Dissertationis/De Lumine/pars prima/ Publice propugnata in Seminario Romano/Societatis Jesu/a Marchione/Andrea Archetti/Academiae redivivorum principe/ejusdem Seminari Convictore,/Augusti XI, Anno MDCCXLVIII//Romae/ Typis Antonii de Rubeis in via Seminari/Romani prope Rutundam./Superiorm permissu.

L. Guzzardi, *Ruggiero Boscovich's Theory of Natural Philosophy*, Science Networks. Historical Studies 60, https://doi.org/10.1007/978-3-030-52093-9_4

> We state that this form of the curve [*hac curvae forma*] explains exceedingly well all general mechanical properties of bodies and many of their special properties; indeed, we advance that they all absolutely depend on this alone [*quin immo censemus omnes prorsus hinc tantum pendere*]. Besides, we can very easily determine the form and immutability of the primary elements of matter, explain mobility, impenetrability, extension, equality of action and reaction (upon which the collision of bodies depends), reciprocal reactions of particles at very short distances (which are quite many and varied), gravity . . ., the cohesion of parts, solidity, fluidity, elasticity and softness, density and rarefaction, and many others as well—including all properties of light. (Boscovich 1748, § 7)

In other words, the mathematical theory of forces, expressed in Boscovich's curve, entails the theory of matter (indeed, this would "absolutely" depend upon that alone). Or, as he would have probably said, it entails its "application to physics." In fact, the above-quoted research program from *De lumine*, duly analyzed in §§ 9–39 of the optical treatise, is recapitulated at the beginning of *Theoria*, Part III ("*Applicatio Theoriae ad Physicam*") as a summary of its contents:

> First, therefore, I will deal with impenetrability, extension, figurability, volume, mass, density, inertia, mobility, continuity of motions, the equality of action and reaction, divisibility, and componibility (for which I substitute infinite divisibility), immutability of the primary elements of matter, gravity, and cohesion. All these are general properties. Then, I will consider the variety of nature, and special properties of bodies; such as, for instance, the manifold variety of particles and masses, solids and fluids, elastic and soft bodies, the principles of chemical operations (solution, precipitation, adhesion and coalescence, fermentation and emission of vapors, fire and the emission of light); also, I will say a few words towards the end about the principal properties of Light, Smell, Taste, Sound, Electricity and Magnetism. (Boscovich 1763, § 359)[2]

Whereas 1748's *De lumine* encapsulated most of the features of Boscovich's research program on the theory of matter, his correspondence with the brothers in that period reveals that, in the spring—that is, shortly *before* writing *De lumine*—Boscovich was working on a more ambitious and far-reaching project. Most of it was accomplished during his regular stays at the Camaldolese hermitage on the Tusculum hill, near Frascati.[3]

[2]A more analytical version of this summary can be found in Guzzardi (2018, 36). It is not my aim in this chapter to give a detailed account of the individual properties in the list. Of course, the "Application of the theory to physics" has important consequences regarding explanations of most of the small and medium range processes that were at the edge of the eighteenth-century experimental research. State changes, sound as well as light emission and propagation, the production and diffusion of heat, chemical processes, electricity, and magnetism—as I will argue, the law of forces for the material points and their aggregates underlie all of these effects. A close analysis of this feature would require a special research all its own. However, my efforts in this chapter will be limited to the exposition of the foundations of Boscovich's conception of matter.

[3]Unfortunately, Boscovich generally refers to such stays with expressions like "Camaldoli" or "Camaldoli near Frascati". This can be puzzling for modern readers, since "Camaldoli" is the toponym of the ancestral seat of the Camaldolese order, which is located in the Casentino Forest on the Tuscan Apennines. But many circumstances in Boscovich's letters, such as other toponyms mentioned in connection with his reference to "Camaldoli", exclude a stay at the original hermitage. To avoid confusion, in the following pages, I will refer to the Camaldolese hermitage on the Tusculum hill as the "Tusculum".

Originally a monastery of the Camaldolese Congregation of Monte Corona, and established in 1606–1607 by the Camaldolese monk Alessandro Secchi, Cardinal Domenico Passionei choose it as his *bon refuge* in the late 1730s, when he returned to Rome from the Apostolic Nunciature to the Emperor in Vienna. Passionei, who became a cardinal in 1738, was a cultivated man and enthusiastic collector of artworks and books; he was often described as an open-minded Catholic, a protector of Jansenists, and a passionate reader of various enlightened authors listed in the *Index*. Determined to find a residence outside of Rome, sometime between 1736 and 1738, he asked the Camaldolese monks at the Tuscolo for their permission to use a couple of cells, with their own gardens as his personal hermitage. In 1739, he transformed them into a small but comfortable home, richly decorated and inclusive of a small library and antiquities of various kinds and sizes.[4]

Passionei began to invite his favorite *letterati* from Rome and abroad to rest and take part in discussions. Due to an increase in the number of their guests, the monks permitted him to use other cells (which also resulted in discontent on the part of the Camaldolese). By the early 1740s, his *buen retiro* had acquired such a reputation that being invited there became the aim of many savants and grand tourers visiting Italy. It is conceivable that, when, on Christmas 1746, the Roman College *professor matheseos* Ruggiero Boscovich was invited for the first time to Passionei's residence on the Tuscolo (see Boscovich 2012b, 122), this was a recognition of his qualities as a man of letters. From then on, Boscovich was a customary guest, occasionally visiting the Camaldolese hermitage on the Tusculum for brief stays over the year and usually spending holidays, such as Christmas, Carnival, and Easter, there.

As reported in his correspondence of 1748, his stays during that year were particularly fruitful. In a letter to Natale (September 14), he commented that he had resumed his current research, much of which was carried out in Passionei's rooms. Based on this document, the 1748 chronology can be sketched as follows:

- In November 1747, Boscovich returned to Rome after a short vacation in Dalmatia from mid-July until the end of October. He left Italy around July 13 and spent most of his time in his native Ragusa. On October 24, he was on his way back to Zadar and reached the Adriatic Coast of Italy on October 30 (see Boscovich 2012b, 167, 134, 135, 137, respectively).
- Once in Rome, he started to work on the imminent lectures and sought "to finish the dissertation on the flow and ebb of the sea," i.e., a projected second part of *De maris aestu*, though it remained unpublished (Boscovich 2012b, 167). In fact, the manuscript *Dissertationis de Maris Aestu/Pars secunda* is preserved in the Boscovich Archives.[5]
- He replied to some mathematical questions posed, "sometimes by letter, sometimes verbally." Two of them are mentioned: (1) an unidentified letter that asked

[4]On Passionei's biography and his multifarious cultural interests, see Caracciolo (1968) and Nanni (2014). On his "*villa al Tuscolo*", see, in particular, Antinori (2004).

[5]See Ruggero Boscovich Papers at the Bancroft Library (University of California, Berkeley). BANC MSS 72/238 cz: Carton 1, Folder 34–36, Item 32.

about some curves led to "a dissertation of some pages" that he planned to publish "sooner or later" (ibid.). It is not clear whether he succeeded in this aim or if this work is preserved in his mathematical manuscripts.[6] (2) An unidentified amateur of mathematics' expression of doubt regarding algebra caused him to refine previous results, which formed the dissertation on the method for raising polynomials to any power.[7]

- At the Tusculum on Carnival 1748 (thus around February 22–27), he "resumed working on the [second part] of the dissertation on the flow, finding some new things" (ibid.).
- Back in Rome during Lent, he "was pressed to give a reply to a letter of a man who stayed here [*uno che stave qui*], who wanted to have some of my things in a collection he was editing, and I could not decline." This was an important turning point; from then on, it seemed to absorb most of his energies. "The question was," he continued, "about some propositions on the infinite divisibility of matter. In my answer, I went deeper into my indivisible points: I lost all restraint [*non fui più padrone di me*] and went straight into the theory, having pondered over it for years." The answer soon expanded in length and became "longer than six pages; but since many things were too concise and there were very few applications to physics, I made some integrations." The joyful effort finally resulted in a fair-copy draft that was "longer than nine pages" but "did not include any figure, and I decided to add some notes for better explication" (ibid., 167–168). With the notes written, there was too much material for an answer, so he changed his mind: "I dealt with answering for few pages and devised the entire work [*ideai tutta l'opera*]" (ibid., 168).
- He presumably brought his papers to the Tusculum when he left Rome for the Easter holiday; Easter fell on April 14, but he was already there by April 8 (ibid., 150). He continued working on his idea and gave it a more refined structure. Perhaps in those days, or some days before leaving, he became convinced that "it was convenient to begin with a detailed presentation [*premettere molto*] of the principles needed for a good discussion of physics." He also realized that "various theorems [*proposizioni*] of Mechanics were often needed." As time passed, he became increasingly inspired: "I started from the simplest definitions [*presa la materia dalle più semplici definizioni*] and I went on to arrange the fundamentals

[6]Potential candidates from the Ruggero Boscovich Papers at the Bancroft Library, Berkeley, include a manuscript presenting a "*quesito*" (question) on curves, mainly on conchoids, and a Latin manuscript *De curvis*, undated (BANC MSS 72/238 cz: Carton 2, Folder 1–23, Items 79 and 82, respectively). A further candidate could be the double letter "Delle ovali cartesiane" ("On Cartesian ovals"), which Boscovich addressed in 1748 to the mathematician Giambattista Suardi, then published in Suardi (1752, 62-79).

[7]"Metodo di alzare un'infinitinomio a qualunque potenza," *Giornale de' letterati*, dicembre 1747, Art. XXXI, 393–404; "Prima parte delle Riflessioni sul metodo di alzare un infinitinomio a qualunque potenza", *Giornale de' letterati*, gennaio 1748, Art. XII, 84–99; "Parte seconda delle Riflessioni sul metodo di alzare un infinitinomio a qualunque potenza", *Giornale de' letterati*, gennaio 1748, Art. III, 12–27.

[*gli elementi*], which are already in fair copy, where I made a variety of nice and noticeable discoveries. I dealt with the rest for a while, working like crazy [*faticando... come una bestia*] until Easter." This agrees with Boscovich's report in a previous letter to Natale (April 23, 1748): "I have been in Camaldoli [on the Tusculum] on Carnival and during these holidays [i.e., the Easter holidays] since I took up this work, and while there, I was more committed than I was here [*la sù ho applicato più che qui*]. Amongst other things, on Easter Monday, I worked all afternoon."

- Other burdens, however, called for his attention. There were dissertations to prepare for students who, according the *ratio studiorum* of the College, were supposed to defend them in public debates. He also extrapolated something from the handwritten second part of *De maris aestu* to be published in *Giornale de' letterati* ("I wrote a dissertation that appeared in the journals of April").[8] In the following months, time was lacking, and "it happened that [he] wrote down the dissertations [for the defendants] at the very last minute [*col laccio alla gola*]." This should have happened, however, in late July, for, at that time, he had already written to his younger brother Bartolomeo (on July 31) that he had completed the dissertation(s) *De lumine* for the defendants in the previous two weeks or so. This agrees with the September 14 letter to Natale: "I draw them anyway; the last one was defended on September 4." (In fact, from the frontispiece of *De lumine, pars secunda*, this was discussed on September 5.)

Some details that Boscovich provided in his September report to Natale are reminiscent of *De materiae divisibilitate*, a work published in 1757 (in a collection of various essays about physics) but conceived and drafted in 1748. In its preface, Boscovich stated, "I had already written this dissertation in 1748, when I was asked about the infinite divisibility of matter" (Boscovich 1757, 131).[9] Also, the circumstance that there were no figures in his papers at first is in accordance with *De materiae divisibilitate*. This begins with the frank admission that no geometric figures are provided in the treatise "for the sake of those who feel annoyed by the continuous inspection" of them (Boscovich 1757, § 2). However, some other details do not match with the later work, which leads one to suspect that this was only one part of the "work of Camaldoli", as Boscovich (2012b, I, 163) referred to it.[10]

[8]"Soluzione geometrica di un problema spettante l'ora delle alte, e basse Maree, e suo confronto con una soluzione algebraica del medesimo data dal Sig. Daniele Bernoulli," *Giornale de' Letterati*, April 1748, Art. XVII, 130–144.

[9]The extended title reads as follows: *De materiae divisibilitate et Principiis corporum dissertatio: Conscripta jam ab anno 1748 et nunc primum edita*. It was published as the fourth chapter of a book collected by Carlo Giuliani (at the time, Secretary of the Accademia degli Oscuri in Lucca), *Memorie sopra la fisica e istoria naturale di diversi valentuomini*, vol. IV (129–258), Lucca: Vincenzo Giuntini, 1757. As Boscovich remarked in the preface, only the footnotes were added in 1757. In the following, I will univocally refer to any passage from *De materiae divisibilitate* as Boscovich (1757), even if the text mostly traces back to 1748.

[10]"Most of my new system is in the dissertation that is now being printed [that is, *De lumine*, Part II]. The more extended study will be the work of Camaldoli [i.e., the Camaldolese hermitage on the

In a previous letter to Natale (April 23, 1748), he gave a detailed description of its concept, and this vastly exceeded the content of *De materiae divisibilitate*:

> The work that I have in my hands, I believe, will extremely increase that bit of esteem I have within the Letters. The thing lingers more than I had thought. By then, I am afraid that the book will be too bulky and it will be convenient to have two tomes. I think that the Mechanics with its notes, which should be the second part, already finished and polished in Camaldoli [on the Tusculum], will exceed 200 in quarto pages in cursive handwriting. I sent it to Father Jacquier, who says that it will bring to me great esteem, and I don't think he is flattering me. The entire work will include four parts full of deep geometry, plus a preface that will give an insight of the whole to the non-geometer. The first part will expose the rules of philosophizing or principles, partly metaphysical, partly geometrical; here, there will be many sublime things about infinity and the methods of the infinitely smalls. The second part will involve the mechanics, which will be entirely deduced from the easiest definitions and with a method I developed by myself; with it, I have progressed much further than anyone else employing the geometrical method. The third section will deal with the smallest parts of matter, and this—as I argue—can be positively proved to not have continuous extension, but be formed by indivisible points spread out on an infinitely divisible space. I determine the force with which these points are endowed, and although they are all similar, I demonstrate with geometrical rigor the immense variety taking place within their diverse aggregates; this only originates from their different disposition. The whole foundation of the fourth section is here; in it, I will prove how, starting from particles of this kind, masses should be formed, which have the same general properties of the bodies in our experience. I apply this theory to thousands of things that follow on their own. The preface, then, presents the history of the different conceptions and philosophical schools, as well as the beginnings of the best discoveries. It also shows that the only progress made has been made where one has walked the same path I'm always walking, so when this path has been abandoned, nothing more has been discovered. Finally, the preface includes a summary of the four following sections. As you see, the work is rich enough [*l'opera è di machina*]. At the beginning, I thought I would simply answer a letter, then I made an 8-page dissertation, then I recast it again, and finally I got completely involved in this whole work . . . (Boscovich 2012b, 155–156)

First of all, the book length does not match with that of *De materiae divisibilitate*, a dissertation that is much shorter than the handwritten bulk of "200 in quarto pages" described above. As for the contents, there are at least two circumstances that are not compatible with *De materiae divisibilitate*: (1) contrary to the initial answer on the indivisibility of matter, geometry would play a chief role here; and (2) the original project was structured in four distinct parts: geometric-metaphysical rules of philosophizing, mechanics, the theory of material points and their forces, and the theory of (macroscopic) matter. *De materiae divisibilitate* only seems to partially cover the third and fourth parts (even if there are overlaps with other parts), whereas the 1758 *Theoria* develops all four topics, even if the architecture of the work was later changed to a tripartite structure with further additions in the supplements.

All of this suggests, on the one hand, that the Camaldolese period was the seminal time in which Boscovich's natural philosophy, essentially based on a Newtonian theory of forces, acquired the strokes of an overall project that culminated in the

---

Tusculum hill]." ["*Il mio nuovo sistema in grandissima parte sta in questa dissertaz:, che si stampa ora. L'o[pe]ra più voluminosa sarà il lavoro di Camaldoli*"] (R. Boscovich to Natale, August 21, 1748, in Boscovich 2012b, I, 163).

Theoria, and I think we have enough evidence to consider the above-quoted outline and the lost manuscript a sort of *Ur-Theorie*. In the subsequent decade, the form of the project would change, though not its fundamental traits. On the other hand, "the work of Camaldoli", initiated by a question about the infinite divisibility of matter, had provoked him to go deeper into the theory of matter, thus establishing *De materiae divisibilitate* as one of the most significant works of those years.

4.2 Deeper into the Points, Building Up Matter

The context revealed in the previous section shows that, for Boscovich, to develop a theory of matter from his theory of force meant to face, and try to resolve, pre-existing tensions within the Society of Jesus. Summarizing some issues that can now be taken for granted, the curve of forces was first conceived as a device to avoid instantaneous changes in the state of motions of bodies; it offered a sort of generalization of Newton's view for both attractive and repulsive forces; it applied to the "particles of bodies", then to larger clusters. But it also suggested—and this was a quite sensitive topic for a Jesuit—that larger, heterogeneous aggregates are composed of smaller, "perfectly homogeneous particles," which were ultimately identified with points.[11] In the 1745 treatise *De viribus vivis*, Boscovich did not investigate further the structure of matter and the features of its elementary components. However, the few details provided could drive anyone trained on Aristotle's texts to suspect that he was cheerfully infringing upon a tenet as important as the infinite divisibility of matter.[12] After all, if he was seriously claiming a *finite* divisibility of matter, did not that claim entail some heterodox atomistic views?

[11]"In all bodies, Newton found a mutual gravity and acknowledged that, in every particle, it diminishes according to the inverse square of the distances. We acknowledge, on the other hand, those repulsions at minimal distances that we mentioned earlier. They grow at infinity as distance decreases. If we only deal with the forces acting on the particles of bodies, they can be represented as follows …" (Boscovich 1745, § 50). "This idea leads us to a composition of larger particles by smaller ones; they are homogeneous, nevertheless the composition is extremely varied" [*Haec autem idea nos perducit ad compositionem particularum maiorum ex minoribus omnino homogeneis, dissimillimam tamen*]" (ibid., § 59). "Bodies are finally revealed as being composed of points." (ibid., § 61).

[12]In the 3 years that followed the publication of *De viribus vivis* (1745), Boscovich's conception raised criticisms by some confreres (see Baldini 2006, 407–409, 432–433). In one case at least, a dissertation discussed in 1746 "denie[d] that the constituents of bodies are 'physical inextended points.' but considers it possible that impenetrability is produced by a '*vis repulsiva*'" (ibid., 432 n 20).

4.2.1 Zeno's Revival

Some updated forms of the atomistic doctrines, frequently found under the appellation of Zenonism, were a highly debated theme within the Jesuit order and raised ambivalent reactions throughout the seventeenth century. In particular, modern Zenonists maintained that the continuum was composed of points and that point-like particles were the ultimate constituents of finitely divisible bodies. Of course, it is not clear whether the person who asked Boscovich about the divisibility of matter was one of his confreres or not; it is also uncertain whether this unidentified scholar saw in Boscovich's theory a dangerous deviation from any Aristotelian orthodoxy or, rather, sympathetically interpreted it as a potential and fruitful revival of atomism. In fact, whatever the questioner's faith, no matter if purely Aristotelian or atomistic of any sort, he might well have expected that Boscovich had clarified the implications of his conception regarding the theory of matter and its relationship with such not entirely conventional, but also not entirely unusual, tendencies.

The standard, official Aristotelianism of the Society of Jesus offered some weapon against atomistic revivals. According to Aristotle, atomists argued that to admit an endless divisibility of matter would lead to aporetic conclusions. As such, let us assume that a body can be divisible through and through:

> Since, therefore, the body is like this everywhere, let it have been divided. What magnitude will be left, then? There cannot be one, for then, there will be something undivided, but it was said to be divisible everywhere. On the other hand, if there is going to be no body or magnitude left, but the division is going to exist, either the body will consist of points and its components be sizeless, or they will be nothing at all, with the consequence that it could come to be and be composed from nothing, and the whole thing would be a mere appearance. Similarly, even if it consists of points, there will be no quantity. For when the points were in contact and there was a single magnitude and they were together, they did not make the whole thing any bigger; for when the magnitude was divided into two or more, the whole was no smaller or bigger than before; hence, even if they are all put together, they will produce no magnitude.[13]

Therefore, atomists concluded, matter must be composed of indivisibles (*atomoi*), i.e., of indivisible primary elements.

Aristotle replied that the assumption of indivisibles would, in turn, equally produce "impossible consequences" and that the atomists' argument actually "conceals a faulty inference." Atomists fallaciously conceive magnitudes as being composed of adjacent (i.e., contiguous) points; however, as in the case of the analysis of the continuous quantities, points are only boundaries of a division procedure. This applies, in particular, to the division of a line, which is described as "a continuous quantity, for it is possible to find a common boundary at which its parts join together, [namely] a point" (whereas in the case of the plane, the common boundary is a line, and in the case of a solid, it is either a plane or a line: see Aristotle, *Categories*, 6, 5a1-6; English version in Aristotle 1984, 8). Hence, when iterating the division of

---

[13]The passage is taken from Aristotle, *On Generation and Corruption*, I 2, 316a 23-34. I employ here the version by Sedley (2004, 69), corresponding to Aristotle (1984, I, 516).

a line, we only find smaller and smaller segments, and no matter how small, each of them is bounded by a couple of points. But Aristotle's key idea is that—as cleared in *De generatione and corruptione*—"no point is adjacent [contiguous] to a point": between any two divisions, a further division is always possible.[14] Thus, the division process will not end up with points, because the process itself is endless. Similarly, when we arbitrarily stop dividing any chunk of matter, we find a smaller chunk, no matter how small: something extended, something not unextended like a point; something, and not nothing.

This Aristotelian background was already present in Boscovich (1745, § 61). Here, he argued that an important parallel exists between mathematical points and points as the basic constituents of bodies. In both cases, points are conceived as indivisible boundaries. Therefore, "as in geometry, mathematical points do not compose a line, a surface, or a continuous solid, but they either coincide with or are separated by any segment; so the physical, real points, as endowed with those forces, cannot compose a continuous extension, but either they must compenetrate ... or they must be separated by any interval."

A more detailed explication, explicitly using the Aristotelian argument from contiguity against Zenonist philosophers, is provided in *De materiae divisibilitate*:

> [Points of matter] are completely unsuitable for composing a continuous quantity. I am brought to this conclusion from the very notion of them, as well as of the continuous extension, so that any two of them cannot possibly be contiguous, but either some interval always lies between them or, if such interval is null, they coincide [*prorsus congruant*] and compenetrate. It is evident from here how much this kind of point differs from the Zenonistic. They are, indeed, regarded as unextended and contiguous to one another, so that they build an extension. [Zenonists' points] are proved to be completely impossible through mathematical demonstrations, as well as clearly appearing absurd at first glance. (Boscovich 1757, § 12)[15]

[14]"Since no point is adjacent to another point, there is one way in which divisibility everywhere is a property of magnitudes, but another way in which it is not. When this divisibility is posited, it seems that there is a point anywhere and everywhere, with the result that the magnitude is necessarily divided into nothing, because the result of there being a point everywhere is that it consists either of contacts or of points. The sense in which it is a property of the magnitude everywhere is that, anywhere at all, there is one point, and all the points are like each other. But there is no more than one in sequence (for points are not <in sequence>), and hence it does not belong everywhere. For, if it is divisible in the middle, will it also be divisible at an adjacent point? No, because no point is adjacent to another point, and it is a point that serves as a division or a join." (Aristotle, *GC* I 2, 316b34-317a12; version by Sedley 2004, 77–78, corresponding to Aristotle 1984, 517). Sedley (2004) also provides a comment and the contextualization of these passages.

[15]This argument, which may be reminiscent of the discussion of continuity and contiguity in Aristotle's *Physics* VI.1, is also exposed in Boscovich (1754b, § 10): "It follows from the nature of boundary itself that a boundary cannot be contiguous to another boundary, for, [given two boundaries,] a continuum, of which they are boundaries, must always lie between them ... One gets the same conclusion even more clearly starting from the indivisibility of a boundary. Either indivisible things are separated from one another or, if the distance is removed [*distantia sublata*], they merge into one thing [*in unicum coalescunt*]. For, either things with no extension do not touch one another or they touch one another according to their totality [*se continugunt sceundum se tota*]. In the first case, they are apart from one another, in the second case, they compenetrate and

There are no direct references to or quotations from those "Zenonists" throughout Boscovich's texts. Since the seventeenth century, an updated version of the doctrine of Zeno had found supporters even among the philosophers of the Roman College, but Boscovich, although aware of these developments, probably used such a label with a rather wide scope, including ancient atomism and Zenonism, as well as its early modern variants. A hint at these can certainly be found in a passage from *De continuitatis lege* (Boscovich 1754b, § 26). Here, he blamed some unnamed philosopher who taught that "an indivisible and very simple particle of matter is extended in length, width, and depth, i.e., it is extended throughout a divisible space so that it occupies the space that ten or one hundred simple and similar particles, but endowed with less extension, could occupy. Some peripatetics called this a virtual extension or divisibility, and some of them have even believed that the same particle occupied sometimes more, sometimes less space, so they dubbed such entities 'inflated points' [*puncta inflata*]."

In fact, Boscovich hinted here at a debate on the nature of the continuum that preoccupied many Jesuit philosophers and theologians during the central decades of the seventeenth century. In particular, the virtual divisibility of space and inflated points played an important role in the discussion of the continuum included in Rodrigo de Arriaga's widespread 1632 textbook *Cursus philosophicus* and was one of the crucial elements of the 1656 *Treatise on the Composition of the Continuum* by Juan de Lugo. The latter was credited by Pierre Bayle (in the *Dictionnaire historique et critique*, article "Lugo, Jean de": see Bayle 1740, III, 219) to have introduced the very notion of a "*punctum inflatum*". This doctrine must trace back to de Lugo's early activity in Spain during the 1610s, but achieved a larger resonance within the Roman College after he was summoned to Rome by the General of the Jesuits, after which it was further disseminated by his pupil Sforza Pallavicino. However, its widespread propagation and the sympathy it enjoyed from highly recognized Jesuit scholars (both de Lugo and Pallavicino had been made cardinals in, respectively, 1643 and 1657) did not save the doctrine of the inflated points from being included in a list, drawn up in 1650–51, of "*Propositiones, quae in scholiis societatis non sunt docendae*".[16]

merge into one thing. For somebody to imagine a point as being contiguous to another point and nevertheless located outside of it and not compenetrated would require a very forced argument. That person will imagine certain little, extended spheres that touch one another on one side and are apart from one another on the other side, thus assuming that the same point has parts and destroying, with that, its indivisibility and inextension. This is, indeed, the oldest argument by which Zeno's conception, that an extended continuum is composed of points with no extension at all, has always been rejected."

[16]On de Lugo's contribution to the continuum debate, see Knebel (2000, 179–180), and Gómez Camacho (2004); I owe a debt of gratitude to Ugo Baldini for drawing my attention to this important background, particularly regarding de Lugo's role and the important contribution of Gómez Camacho. On the inclusion of Zenonist doctrines in the mid-seventeenth century list of "propositions not to be taught in the Jesuit schools," see Festa (1992, 203, 1999, 107–108), Palmerino (2003, esp. 187–205), Gómez Camacho (2004, 30–31), and Knebel (2011, 200–201). On Arriaga, his potential influence on Boscovich, and the role of Zenonism in shaping his attitude toward

All of this may contribute to explaining Boscovich's attitude toward such doctrines and why he considered Leibnizian monadology as one variant of the Zenonist menace, in so far as Leibnizians are charged to maintain that "monads are unextended, but in such a manner that they … compose an extended continuous quantity"—a conception that causes them to fall "into the same absurdities in which Zeno's points also fell" (Boscovich 1757, § 12). To conclude, a footnote added to these passages in 1757 points out that monadology would also face the argument from contiguity that was "established against Zenonists many centuries ago. And this argument has never been satisfactorily answered" (Boscovich 1757, footnote *a* to § 12, p. 153; the passage was then extrapolated and included in Boscovich 1763, § 139).

4.2.2 *Aristotelianism Corrected with Newtonian Transduction*

Of course, the rebuttal of old and new forms of atomism was only one of the multiple facets involved in Aristotle's texts on continuity. The analysis of the continuous quantity in terms of an infinite divisibility of parts had a distinct aspect that a mathematical reader of the eighteenth century could hardly ignore. A geometric quantity—a line, for example—can be thought of as infinitely divisible, so that one may obtain infinitesimal small parts. But how should we consider the infinitely small quantities arising from this procedure? Should we think of them as vanishing magnitudes, as "ghosts of departed quantities", as George Berkeley famously did, or are they determined quantities that are, by their essence, actually infinitely small? So, for an eighteenth-century Jesuit mathematician, Aristotle's text concealed the question of the status of the infinitesimals. It may not have implied a solution, but it certainly provided an orientation.

The significance and justification of "fixed" infinitesimal quantities had preoccupied generations of mathematicians since the late eighteenth century, beginning with Newton, Leibniz, and their controversy. Boscovich entered the debate in 1741, with a work that investigated the correct use of the concept of infinity and infinitely small quantities as a consequence of their "nature", for "unless the nature of infinite quantities is carefully examined, their utility is compromised by danger of error, whether we use them in analysis or geometry" (Boscovich 1741b, § 1). After

monadology, see Rossi (1999, 76–91). On Zenonism and its reception, particularly in relation to Leibniz, see Beeley (1996, 298–300). Fernandes (2013, esp. 470) has suggested that Giambattista Vico might have played a role as a potential source for Boscovich's stance toward Zenonism. In light of the previous reconstruction, I cannot see any evidence of Vico's influence; I rather think that, regarding Zenonism, Boscovich and Vico had some sources in common.

rejecting the concept of the indivisibles, he claimed that "there are no constant infinitely small quantities that are determinate."[17] The proof reads as follows:

> If a straight line segment could be divided into infinitely many equal parts, the infinite would become [*evadere*] finite by removal of a unit subset. For, between the first and last parts would be infinitely many parts, and similarly between the second, third, or tenth, and the very last. But there will not be infinitely many between the last one and the one removed from it by ten or twenty parts. If we now suppose all and only those removed that have infinitely many parts between them and the last one, then, when the final one of them has been removed, necessarily, the remaining ones are infinitely many; for, otherwise, the final one removed would not have been. If we do not consider the first remaining [*at non computata reliquarum prima*], their number will be finite; otherwise, we would have had to remove it. (ibid., § 10)[18]

In fact, Boscovich claimed that both infinity and infinitely small quantities are indeterminate. They are considered as quantities "either diminished or increased beyond arbitrary bounds." Specifically, "a quantity taken to decrease beyond arbitrary bounds, we term *infinitely small*; one taken to increase beyond arbitrary bounds, *infinite*" (ibid., § 12).

In other words, according to Boscovich, geometric continuity is obviously consubstantial with the idea that an extended, *geometric* quantity can be infinitely divided. This only means, however, that the number of potential divisions can be increased beyond any arbitrary bound—and, correspondingly, that the length of the resulting subsections can be diminished beyond any arbitrary bound. However, this does not entail the idea that the line is composed of an *actual* infinity of subsections. Rather, a person may cut a line however many times she wants; still, after *any* number of cuts, she is left with a *finite* number of them. But, of course, after a cut that one might have considered as the "final" one, she is nonetheless allowed to perform a new cut, if she wants, and so on. In this sense, a person has at her disposal a *potential* infinity of cuts, but after each step, she may count how many cuts she has made—so they are actually finite in number.

Now, let us remember that, by cutting, viz. dividing, a line, an individual creates intervals included between boundaries. In other words, the person is creating points, for points are boundaries of closed linear intervals. How many points can someone create by cutting the resulting intervals through and through? Potentially, she can

[17]Note that Boscovich directly or indirectly resumed this characterization in later works, always referring to his (1741b). See, for example, Boscovich (1754b, § 80): "To be sure, in the dissertation *De natura, et usu infinitorum, et infinite parvorum*, issued many years ago, we proved [*demonstravimus*] that there cannot possibly be infinitesimal quantities determinate in themselves ..."; Boscovich (1763, § 90): "I proved clearly enough, I think, in the dissertation *De natura, et usu infinitorum ac infinite parvorum* ..., that there are no infinitesimal quantities determinate in themselves."

[18]Here, I do not question the strength of this proof, which is incomplete and unsatisfying in view of modern standards. As Homann (1993, 414) remarked, this argument relies on a visualization technique and lacks geometric, as well as logical, rigor; for example, we are not guaranteed that the elements chosen should have first and last elements, as Boscovich's proof requires. In the quoted passage, I follow Homann's version (ibid., 426), with some variations.

create an infinity of them. Therefore, in geometric continuity, the infinite divisibility of space and the endless insertibility of boundaries (of points, in the simplest case of the line) are essentially the same: Given any two boundaries, one can interpolate how many points she wants by "cutting the line". For many reasons, this is clearly an Aristotelian strategy.[19] Note, in particular, that any potential point can be brought to actuality. Moreover, the intervals obtained are adjacent, because each of them, Aristotelically, has a common boundary with its successor.

Nevertheless, this strategy also has its limits. According to Aristotle, there is no material difference between physical and geometric entities, since the mathematician investigates the same objects as the physicist, even if their modes of investigation are different.[20] Matter is infinitely divisible, just as any geometric entity is. According to Boscovich, however, a subtler parallel must hold. As in the case of geometric space, if we divide matter through and through, we create intervals enclosed by boundaries; but such intervals are filled with matter, whose parts, as we have seen, are endowed with forces acting at a distance. Now, there is no reason why this property should cease to be in effect as we reduce the scale of matter under consideration. More precisely, Boscovich thinks that *there is* a reason why this extension of properties from the macrolevel to any microlevel should be taken for granted: namely, the "analogy of nature".

This expression occurs many times and in different contexts throughout Boscovich's works. As remarked upon by Martinović (1987, esp. 95–96), it is reminiscent of an assertion that Newton (1687, 795) made in his commentary to the Third Rule of philosophizing: "Nor should we depart from the analogy of nature, since nature is always simple and ever consonant with itself." Now, this rule prescribed that "those qualities of bodies that cannot be intended and remitted [i.e., qualities that cannot be increased and diminished] and that belong to all bodies on which experiments can be made should be taken as qualities of all bodies universally." In the commentary, before his endorsement of the analogy and simplicity of nature, Newton also explained that "the qualities of bodies can be known only

[19] For comparison, see the "operational" method with which Aristotle explained continuity, as described by Ugaglia (2012, 26–27).

[20] "The next point to consider is how the mathematician differs from the student of nature; for natural bodies contain surfaces and volumes, lines and points, and these are the subject-matter of mathematics .... Now the mathematician, though he too treats of these things, nevertheless does not treat of them as the limits of a natural body .... Similar evidence is supplied by the more natural of the branches of mathematics, such as optics, harmonics, and astronomy. These are in the way the converse of geometry. While geometry investigates natural lines but not *qua* natural, optics investigates mathematical lines, but *qua* natural, not *qua* mathematical" (Arist., *Ph* II 2, 193b23-194a12). "Just as the universal part of mathematics deals not with objects that exist separately, apart from magnitudes and from numbers, but with magnitudes and numbers, not however *qua* such as to have magnitude or to be divisible, clearly it is possible that there should also be both formulae and demonstrations about sensible magnitudes, not however *qua* sensible but *qua* possessed of certain definite qualities" (Arist., *Metaph* M 3, 1077b18-23). See Ugaglia (2012, 35–37). On the peculiar ontological status of the mathematical objects according to Aristotle, see Cleary's (1995, 269–344) analysis of Book *M* (XIII) of *Metaphysics*.

through experiments; and therefore qualities that square with experiments universally are to be regarded as universal qualities ... Certainly idle fancies ought not to be fabricated recklessly against the evidence of experiments." Thus, as noted by McGuire (1970, 3), in this context, the analogy of nature functioned as a justificatory principle "by which to sanction inferences from what is observable to what in principle is unobservable"—something that McGuire called, following Maurice Mandelbaum, "the problem of transduction": the problem of how to extend the laws and properties of observable macroscopic bodies to the imperceptible microscopic parts that are supposed to be their components.[21]

In the next chapter, I will argue that Boscovich initially saw in the analogy of nature a principle of higher order, not necessarily limited to the theory of matter—a strategy that he later abandoned. And, of course, it could even be the case that Boscovich picked up the phrase "the analogy of nature" from other sources (to be sure, Newton was not the only one to express this concept!)[22] In any event, an example of transduction, with a transparent reference to the analogy of nature as a guiding principle for the insight of material points, is the following passage from *De materiae divisibilitate*, probably tracing back to the Camaldolese *Ur-Theorie* of 1748, as the use of the term *adhaesio* instead of *cohaesio* would seem to indicate:

[21]More precisely, as Ducheyne (2012, 208) brilliantly explains, "in the context of transduction, the 'Analogy of Nature' was a guiding principle in Newton's research, for without the uniformity of microscopic and macroscopic components, it would be impossible 'to derive the qualities of imperceptible bodies from the qualities of perceptible ones,' as Newton observed in manuscript material on Hypotheses III (later Rule III) in 1692." However, Ducheyne also shows that, in the following years, his growing devotion to optical research and experimentation suggested to him that the use of analogy should be constrained—otherwise, an uncontrolled use of analogic reasoning could lead to some undesirable consequences, as far as it could encourage unscrupulous hypotheses on per se unobservable micro-constituents of the optical phenomena (see also Shapiro 1993, 134). I owe a special thanks to Steffen Ducheyne for having discussed this issue with me and for his suggestions.

[22]Ivica Martinović (1987) was probably the first to appropriately remark upon the importance of the analogy of nature to Boscovich's epistemology in connection with Newton's third rule (but also see Nedelkovitch 1922, 106). However, he failed to recognize its role in arguing for the material points (by the way, contrarily to him, I think that analogy is particularly significant in Boscovich's thought exactly in such a context, not in arguing for the principle of continuity). Ivica regretted being unable to "find any proofs of the direct influence of this view of Newton [about the analogy of nature as expressed in the third rule] on Bošković" (ibid., 96). It seems to me that the evidence at least suffices to argue that Boscovich was aware enough to use a concept endorsed by, amongst others, Newton: after all, Boscovich (1757, 86) quoted the passage from Query 31 where nature is said to be "very conformable to herself and very simple" (Newton 1730, 397). But he probably knew that this concept had other incarnations before Newton (it was shared, for example, by Descartes, Boyle, and many others: see Mandelbaum 1964, 61–117; Shapiro 1993, 40–48). And so, Boscovich could enlist Newton among those who were committed to analogical reasoning, even if he occasionally disentangled the analogy of nature and Rule III: see, e.g., Boscovich (1755b, 46) footnotes *a* and *b*, where Rule III is said to "include the principle of induction [*principium inductionis continent*]," whereas Rule I and II are considered as "depending on a certain simplicity and analogy of nature [*utraque regula pendet a simplicitate quondam, et analogia Naturae*]." As we shall see in the next chapter, the extent to which he believed that the principle of analogy was applicable remains questionable.

> Since, at greater distances, there is, as is well known, at least that attraction that is due to the universal gravity, there must be a limit point at which, as distances are increased, repulsion passes into attraction. It is shown that this limit point is a limit of adhesion [*limitem adhaesionis*], so that particles posed on it must observe the distance that they got once. The adhesion of particles composing bodies should be considered as consisting of this kind of limit point. *The analogy of this and the simplicity of nature allow for extending this idea of adhesion to every adhesion* in which a greater particle results from the composition of smaller ones ... *That same analogy let exclude another kind of adhesion* ... *This analogy drove me to the indivisible points.* (Boscovich 1757, § 90; emphasis added)[23]

Let us try to make explicit Boscovich's analogic-transductive reasoning: we can imagine reducing intervals, but, according to the "analogy of nature", any interval of matter, however small, will preserve inertia, as well as the determinations to approach one another or recede from one another depending on their reciprocal distances. Now, if we imagine dividing any extended material object, its constituent parts cannot be thought of as adjacent, because the law of forces would keep their boundaries at some distance, since repulsion grows infinitely when distance is infinitely diminished. Of course, an infinitely small distance is not a definite, assigned measure; rather, it is shorter than any assigned measure—but it is certainly a distance. And so, if we indefinitely iterate the process of dividing a material body, reducing the intervals in every direction, we shall ultimately end up with the material boundaries, or points, of such an infinitesimal distance. Points are material, but that distance must necessarily be void of matter, for if there were to be matter, then the points forming it would be subject to the law of forces again, causing them to

---

[23]Boscovich also used the argument from the analogy of nature in answering possible objections. Let us compare the above-quoted passage with a couple of paragraphs from the *Theoria* (note that, here, the term 'cohesion' is employed: *adhaesio* occurs in the 1745 *De viribus vivis*, whereas *cohaesio partium* is first used in Boscovich 1748, §§ 20 ff., and largely preferred in Boscovich 1754b, 1755a): "[81] Because the repulsive force is infinitely increased when the distances are infinitely diminished, it is quite easy to see clearly that no part of matter can be contiguous to any other part; for the repulsive force would at once separate one from the other. Therefore, it necessarily follows that the primary elements of matter are perfectly simple, and that they are not composed of any parts contiguous to one another. This is an immediate and necessary deduction from the constitution of the forces, which, at very small distances, are repulsive and increase infinitely. [82] Perhaps someone will here raise the objection that it may be that the primary particles of matter are composite, but that they cannot be disintegrated by any force in nature; that one whole with regard to another whole [*quarum altera tota respect alterius totius*] may possibly have those forces that are repulsive at very small distances, whilst any one part with regard to any other part of the same particle may not only have no repulsive force, but indeed may have a very great attractive force such as is required for cohesion [*cohaesionem*] of this sort ... But, in the first place, this would be in opposition to the homogeneity of matter, which we will consider later; for the same part of matter, at the same distances with regard to those very few parts, along with which it makes up the particle, would have a repulsive force; but it would have an attractive force with regard to all others, at the very same distances; and *this is in opposition to analogy* ... Lastly, with this idea, there would have to be two kinds of cohesion in Nature that were altogether different in constitution; one due to attraction at very small distances, and the other coming about in a far different way in the case of masses of elementary particles, that is to say, due to the limit points of cohesion. Thus, *a theory would result that is far less simple and less uniform than mine*" (Boscovich 1763, §§ 81–82; emphasis added).

separate, etc. Moreover, since an extended material object by definition occupies a finite volume (has a finite extension), the set of the distances separating its constituent points is finite, causing the number of points itself to be finite as well. As a consequence, matter is only finitely divisible.[24]

In sum, when Boscovich argues that the ultimate constituents of matter are its points, he implicitly appeals to a process of division of bodies and to an argument from transduction. From the Aristotelian tradition, he took the idea that material points are boundaries that enclose any interval of matter, however small; being boundaries, they are obviously endowed with geometric properties: they are "perfectly indivisible and unextended" (Boscovich 1763, § 7). Being material, they are also endowed with—*via* the analogy of nature—two basic non-geometric properties coming from his theory of forces: inertia and the determinations to approach one another or recede from one another according to their reciprocal distances.[25] And this causes them to be at a distance.

4.3 Never-Ending Aggregates

The now established concept of the material point drove Boscovich deeper into the fine-grained structure of matter. It also signals a radicalization of his program, for he emphasized that the curve of forces primarily applies to material points, thus governing the internal structure of bodies, as well as their mutual relations. The final result was a sophisticated mechanism, mostly exposed in the third part of the *Theoria*, that allowed for extending the mechanical properties of points to aggregates, which can be categorized into different orders. The most fundamental properties, such as the impenetrability of the primary elements of matter, extension, shape, and volume, can easily be derived from the dynamics of a system of points (Boscovich 1763, §§ 360, 371–372, 375–377); other physical properties are then obtained by applying the law of forces to those larger aggregates.

First, let us remember that limit-points are balance points between attraction and repulsion, where positive and negative accelerations exactly compensate each other, resulting in a zero on the abscissa. However, as I briefly discussed at the end of Sect. 2.4, not all limit points are created equal—not all of them have the same meaning. While examining Boscovich's curve (Fig. 2, Introduction, reproduced here below), let us assume that A, the origin of the axes, is the position of a fixed point that we use

[24]"If the primary elements of matter are perfectly non-extended and indivisible points, separated from one another by some definite interval, then the number of points in any given mass must be finite; because all the distances are finite" (Boscovich 1763, § 89).

[25]See, in particular, Boscovich (1745, § 61, 1748, II, § 8, 1757, § 18, 1754b, §§ 165–166, 1763, §§ 8–9, 136). In the following, I will use the expressions 'material points' and 'points of matter' interchangeably.

to appreciate the motion of a material point α. In general, we may have three different cases. First, consider the two extreme positions:

1. when α is very near A, it tends to accelerate toward E; at the beginning, the acceleration is greater, but as α goes away, it lessens;
2. when α is very far away, such as at planetary distances, it moves toward T, S, and R, approximating Newton's inverse-square law. Now, the situation is more or less the opposite of (1): When α is very far away, the motion is slow but accelerates as it approaches T and S (and then slows down again while approaching R).

The interesting part is what happens, so to speak, "in between" the extremes, as soon as we consider a third possibility:

3. α arrives at intermediate points, such as E and G. Both are zeros on the x-axis, meaning that they are points of "indifference" (which Boscovich calls "limits"), where positive and negative accelerations compensate each other. But, in a certain sense, they are endowed with different properties. Let us imagine that α is in G. What if α gets determined by any external action (such as a point β) to move toward A through the position E? It will tend to easily acquire the new position E, because its acceleration grows toward F; then, it slows down but does not invert its sign until it reaches E. But now, once it reaches E, α can hardly continue moving forward. If it were determined by any external action to approach A, it will easily tend to recover the previous position, E, due to the action exerted by A itself. In contrast, let β determine α to move away from E; then, a contrary acceleration determined by A on α itself will take place, for which α will tend to move toward E again. But, if α were to hardly reach G, and β keeps on determining it to move away, toward I, L, N..., α will easily perform the motion in this direction (Fig. 4.1).

This last case is particularly relevant to the theory of matter, since it allows us to distinguish two different classes of limit points, or intersections between the curve and the x-axis, that may produce different stability conditions of compounds: E-type intersections, "which tend to preserve the mutual position [and are called] limit points of cohesion," and G-type intersections, i.e., "limit points of non-cohesion" (Boscovich 1763, § 180). Thus, the curve can be viewed as a succession of E-type and G-type intersections with the x-axis.[26]

As can easily be seen, the strength of such conditions depends on how quickly the curve departs from the axis at the point of intersection—in modern, analytical terms, it depends on the derivative of the force function at the point of intersection, or the slope of the tangent line to the curve at that point. So, as regards the E-type intersections, "there can be very strong, as well as very weak, limit points of

[26]"Since, at one intersection, the curve passes from the repulsive part to the attractive part, at the next intersection, it is bound to pass from the attractive to the repulsive part, and vice versa. It is clear then that the limit points will be alternately of the first and second kinds" (Boscovich 1763, § 179). "The points of intersection are alternately limit-points of non-cohesion and cohesion" (ibid. § 411).

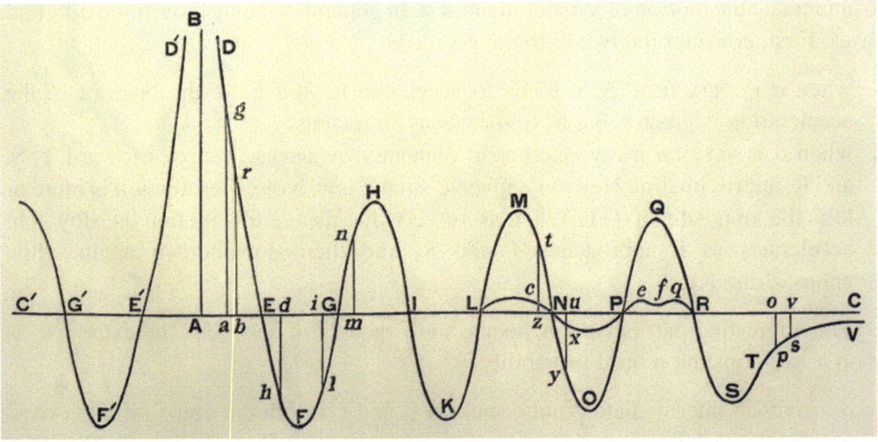

Fig. 4.1 Boscovich's curve. (From Ruggiero Giuseppe Boscovich, *A Theory of Natural Philosophy*, ed. by J.M. Child, New York: The Open Court, 1922). In truth, this is just a basic instantiation out of many possible shapes, as can easily be seen from the analytical expression of the force function that Boscovich published in one of the "Supplements" to the treatise (see *this book*, Sects. 4.3, 6.4, and 6.5)

cohesion. If the curve cuts the axis almost at right angles and goes off far away from it," that is, if the shape of the curve near that point is steep (or the value of the derivative of the curve at that point is great), "then they are very strong. But if it cuts the axis at a very small angle and separates from it but little, then they will be very weak." A proof in quasi-analytical, although informal, terms is then provided:

> The arc N*yx* in Fig. 2 represents the first kind of limit points of cohesion, and the arc *c*N*x* the second kind. At point N, let N*z*, N*u* be assumed, however small, to be on the *x*-axis; then, the force intensities [*vires*] *zt*, *uy*, as well as the areas N*zt*, Ny, may be as great as we please. Thus, if distances are changed of a quantity, however small, there can be found force intensities, however great, expressed by the ordinates, which will strongly resist a compressing or separating force, be it as great as we please; there can also be found areas, however large, that will annihilate the corresponding velocities, however great. Hence, a sensible change in the mutual position [of the material points] can be hindered, in opposition to any impressed force, however great, or against a velocity generated by the actions of other points. In the second kind of limit-points of cohesion, even if the segments N*z*, N*u* are taken to be of considerable size, both the force intensities *zc*, *ux* and the areas N*zc*, N*ux* can result as small as we please; therefore, the resistance preventing the change will likewise be as small as we please. (Boscovich 1763, § 182)

Points that are at distances of limit points of cohesion tend to be in a stable (or metastable) state. A symmetric situation occurs in the case of limit points of non-cohesion: here, the most stable condition occurs when the derivative of the force function is arbitrarily near to zero at G-type intersections. In this case, indeed, a material point α near G is attracted toward E (or repelled toward I) by a force intensity, however small; it cannot recover position G on its own, but, if the derivative is arbitrarily near to zero, the displacement will be arbitrarily slow.

Another symmetry between limit points of cohesion and non-cohesion involves oscillations. If the force impressed upon a material point α near a G-type intersection is enough that α is moved toward the following (weak) E-type intersection, then an oscillation can take place (of course, E-type intersections may also involve oscillations, although possibly smaller). In particular:

> If any force separates [two material points] at the typical distance of a [E-type intersection] . . ., then the oscillation will be exceedingly small, at least so long as the limit point is a fairly strong one. For the velocity will immediately begin to be diminished, a force contrary to that one will be found at once, and the points, being moved but little from their original position, will immediately afterwards retrace their paths if left to themselves. But if they are separated by a distance equal to that of a [G-type intersection], by any force, however small, then the oscillation will be much greater; for they are necessarily bound to go on beyond the distance equal to that of the next [E-type intersection], and the motion will not begin to be retarded until this has been done. (ibid., § 194)

In Part II of the *Theoria*, Boscovich described other mechanical features of material points aggregates that it is not my aim to present in detail. He also claimed that the alternation of E-type with G-type intersections (including their further specifications: see ibid., §§ 195–198) underlies many phenomena that will be dealt with in the "Application of the theory to physics" (Part III), from fermentation to the emission and propagation of light, elasticity, etc.[27] Of course, this mechanism also provides the foundation of (meta)stable compounds and explains the force that holds bodies together—the force that Newton called cohesion—as an effect of material points being at distances typical for E-type intersections.[28] As we have seen, in the case of a strong limit of cohesion, "the [material] points will preserve such a distance with the greatest strength, so that, if they are imperceptibly compressed, they will resist further compression, and, when pulled apart, they resist further separation." Of course, this also applies to larger aggregates of points—if these "cohere together [*cohaereant inter se*], they certainly will preserve their positions and form a very stable mass as regards its shape [*massam constituent formae tenacissimam*], which will display exactly the same phenomena as little solid masses, as commonly understood" (Boscovich 1763, §§ 165).

Let us now attempt to sketch the formation process of bodies: as we shall see, Boscovich distinguished various orders of particles and aggregates (even if he never

[27] "The great agitation, with its various oscillations and motions that are sometimes accelerated, sometimes retarded, and sometimes reversed, will represent fermentations and conflagrations. The starting forth from a very large repulsive arc with very great velocities, which, as soon as very great distances have been reached, are very little different from one another; nor are they sensibly changed in the slightest degree for very great intervals; this will represent the emission and uniform propagation of light, and the approximately equal velocities in any ray of the same kind from the stars, the sun, and a flame, with a very slight difference between rays of different colours. The force persisting after compression, or separation, will serve to explain elasticity. The lack of motion due to the frequent occurrence of limit-points, without any endeavour towards recovering the original configuration, will suggest the idea of soft bodies" (ibid., § 199). A more detailed description of Boscovich's theory of matter with reference to the stability conditions of the aggregates is provided by Ullmaier (2005, 60–89).

[28] As noted above, there is, unfortunately, some variability in Boscovich's usage of the term cohesion: whereas Newton usually applied it to interparticular interactions, Boscovich (1763) oscillates between the broader but technical meaning of the "*limites cohaesionis*" and the idea of an interparticular force.

provided a description for aggregates formed by more than four points: see, on this, Ullmaier 2005, 75). Imagine having a substantial mass of many scattered points. Some of them will be nearer, resulting in states of strong attraction (hence cohesion), and some others will be far enough away to have less cohesion or even repulsion, which, again, can be counterbalanced by an interplay of other points' attraction and repulsion. So, points first tend to cohere into smaller aggregates; these, in turn, tend to repeatedly cohere into greater ones until they finally form bodies—which can be more precisely thought of as *aggregates of aggregates of points*:

> Now, my elements are really such that neither themselves, nor the law of forces can be changed; and the mode of action when they are grouped together cannot be changed in any way; for they are simple, indivisible and non-extended. From these ..., when collected together at very small distances apart, in sufficiently strong limit-points on the curve of forces, there can be produced primary particles, less tenacious of form than the simple elements, but yet, on account of the extreme closeness of its parts, very tenacious in consequence of the fact that any other particle of the same order will act simultaneously on all the points forming it with almost the same strength, and because the mutual forces are greater than the difference between the forces with which the different points forming it are affected by the other particle. From such particles of the first order, there can be formed particles of a second order, still less tenacious of form, and so on. For the greater the composition, and the larger the distances, the more readily can it come about that the inequality of forces, which alone will disturb the mutual position, begins to be greater than the mutual forces that endeavour to maintain that mutual position, i.e., the form of the particles. (Boscovich 1763, § 398)

The geometrical properties of the aggregates, by means of the system of E-type and G-type interactions, should also explain the differences between particles: they arise "from the number of points in them, their volume, their density, their shape" (ibid, § 419) and result that each body may have (and generally has) an inhomogeneous constitution. Its most basic constituents, the points, are homogeneous, but any other order of particles is not (or not necessarily):

> The points in one particle may be disposed in a sphere, in another in a pyramid, or a square or triangular prism. Take any such figure, & suppose the points are disposed in any particular manner whatever; then, there will be as many distances as there are pairs of points, & their number will be finite in every case. The curve of forces can have any number of limit-points of cohesion, & these can occur anywhere along it. Hence, it must be the case that limit-points can be found to correspond to those distances, on account of which the particle will have that particular form and can be extremely tenacious in keeping that form ... Apart from the fact that to each distance there corresponds a limit-point in the primary curve, or that there are pairs of asymptotes, or any other asymptotes of the sort except the first, there are really an innumerable number of kinds of figures, in which, with a given number of points, there can be equilibrium, & a limit-point of cohesion due to the cancelling of equal & opposite forces. (ibid.)

Therefore, aggregates may differ based on their constituents rather than on the law related to their constitution, which is the same curve of forces that holds for points—only the result is more complex because of the number of points and the great variety of mutual distances potentially involved. Hence, differences among bodies have to be ascribed to the diverse geometric arrangements of their constituent parts. In this manner, what Parts II and III of *Theoria* explained in terms of the fine mechanism of mutual attractive/repulsive forces acting between points now counts as a "property" of an aggregate:

> Back to my theory of homogeneous elements again, the several forms of bodies will consist in the combination of homogeneous points, which comes from their distances and positions, and, in addition to combination alone, the velocity and direction of the motion of each of the points; however, instead of the individual masses of bodies, the total number of points will be assumed. Given the number and disposition of the points in a given mass, the basis of all its properties, which are inherent in the mass, is given; and also that of all specific forms; hence, the relations that the same mass must have with other masses; that is to say, those determined by their numbers, combinations & motions; moreover, the basis of all changes that can happen to it is also given. (Boscovich 1763, § 519)

Gravity is not but a final step: it is another "property of matter" (ibid., § 399), whose emergence depends on the more substantial distances dominating within a super-aggregate (the Universe) of aggregates (the celestial bodies) of aggregates (their parts) of a multiplicity of aggregates (the particles) of points. However, Boscovich also pointed out that Newton's gravity only closely approximates the value obtained by the curve of forces, where it is represented by the last arc; in other words, gravity is not exactly proportional to the inverse square of the distances, but is instead proportional to a function of them.[29] Moreover, it does not expand on every scale, "but merely on distances such as those that lie between the distance of our bodies from the far greatest part of the mass of the Earth, and the distances from the Sun of the aphelia of the most remote comets" (ibid.). Since the forces depend on distances, there are many distances (so to speak, in the short and mid-range) at which gravity does not emerge.

This interplay of forces, which originated from the multiplicity of distances between homogeneous points, was already explained in the 1748-drafted *De materiae divisibilitate*:

> Whenever at certain distances, the law of forces will be such that, if distances are modified as much as the diameters of the particles, the forces will not be sensibly changed (that is to say, whenever a difference of the abscissas equal to the distance between the extremes of the particles would correspond to an unnoticeable difference of the ordinates compared to the ordinate of the whole particle), an equal number of particles will follow, according to our senses [hence, approximately], the same law that individual points would follow ... However, if the change of the distances would cause a considerable variation of forces (i.e., an appreciable change of the ordinate compared to the ordinate of the whole particle), at those distances, the particles will follow a very different law from that followed by individual points ... In the first place, this will certainly affect the universal gravity, which is exerted upon distances compared to which the diameters of the individual particles are extremely small and is such that, if the distance slightly changes, it does not change in any appreciable way ... Second, all of this does affect the cases in which, if the distances are slightly changed, the forces vary to extreme degrees. Here, the limits take place frequently [so there are many limits in a small segment along the abscissa] and the forces are stronger, so that the curve departs at most from the axis. Therefore, the forces that particles exert one upon the other—the forces that occur in the chemical effects, in the formation of metals, as

[29]Of course, Boscovich's gravity is proportional to the product of masses: "Now, the principal laws of gravitation are that it varies directly as the mass and inversely as the square of the distances from each of the points of that mass. In my theory, it is quite clear that this must be the case. For, as soon as we reach the arc of my curve that represents gravitation, all the forces are attractive, and to all intents obey the same law. So, some of them do not cancel out others in opposite directions, but their sum approximately corresponds to the number of points" (Boscovich 1763, § 401).

well as of plants, in the generation of every body from the motion of particles whose reciprocal distances are very small with respect to their diameters—will be very different from the forces that the individual points exert upon other matter and one upon the other. (Boscovich 1757, § 46)[30]

The term *law* in this passage is used in a somewhat confusing way. Of course, Boscovich meant that the law of forces is ubiquitous. However, since points are generally found at very different distances, the individual forms of the curve of forces (i.e., the local "laws") may greatly differ.

* * *

There is a final aspect of the logic of the aggregates that is worth mentioning. Together with the idea of the first, asymptotic arc of the curve that represents repulsion, the concept of a multiplicity of aggregates essentially obeying the same law of forces led Boscovich to speculate about the possibility of a plurality of "worlds" (*mundia*).

As we have just seen, the individual forms of the curve of forces followed by the aggregates become more complex as the aggregates become more complicated. In particular, Boscovich's study of the mathematical underpinnings of the law of forces caused him to also consider curves with multiple intermediate asymptotes, as shown in the following Fig. 4.2 (I will return to this in Chap. 6.)

Boscovich then offered this explanation in Part II of *Theoria*:

If a number of points are assembled between any pair of asymptotes, or between any number of pairs you please, correctly arranged, there can arise any number, so to speak, of worlds [*posset exurgere quivis, ut ita dicam, Mundorum numerus*...]. And this happens in such a way that none of them has any communication with any other, since, indeed, no point can move out of the space included between these two arcs, one repulsive and the other attractive; and such that all the worlds of smaller size [*Mundi minorum dimensionum*] taken together would act merely as a single point towards the next greater world, which would also consist of point-like masses as considered in itself; for, to be sure, every size of the individual [worlds], with regard to that same world and to the distances that they can attain within it, is practically nothing. From this, it could also follow that any one of these worlds would not be appreciably influenced in any way by the motions and forces of a greater world. In any given time, however great, the whole inferior world would experience forces, from any point of matter placed without itself, that approach as near as possible to equal and parallel forces. These, therefore, would have no influence on its relative internal state.[31]

[30]Let us compare this with Boscovich (1763, § 402): "We see so much uniformity in all masses with regard to the force of gravity; in spite of the fact that these same masses, for the purpose of other phenomena depending on the smaller distances apart, have differences so great as those possessed by different bodies as regards hardness, colour, taste, smell, and sound. For, a different combination of the points of matter induces totally different sums for those distances up to which the curve of forces still twists about the axis; where a very slight change in the distances changes attractive forces into repulsive, and substitutes, vice versa, differences for sums. Whereas, at those distances for which gravity obeys the laws that we have stated very approximately, the curve has its ordinates all in the same direction and, even if the distance is slightly altered, practically unaltered in length. This, of necessity, produces a huge difference in the former case, and a very great uniformity in the latter."

[31]Boscovich (1763, § 171). Child's translation is completely misleading in this case. The original version is as follows: "*Collocato quocunque punctorum numero inter binas quascunque*

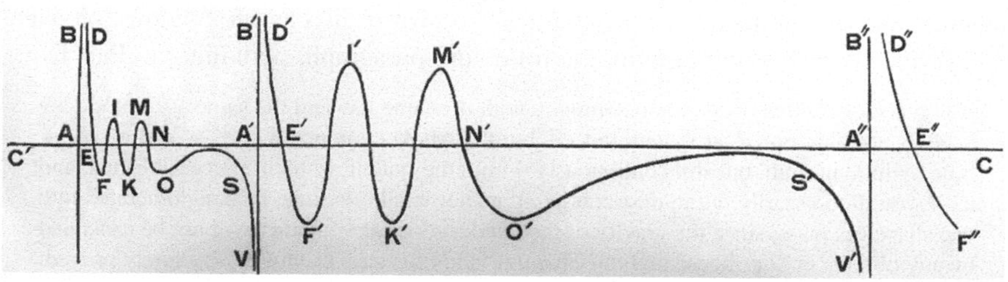

Fig. 4.2 Boscovich's scheme of possible distinct, non-interacting "worlds" (aggregates), from Fig. 14 of Child's 1922 edition of the *Theoria*. Let "any number of segments [be] AA′, A′A″, so that any successor is immensely great compared with any predecessor. Let the asymptotes AB, A′B′, A″B″, perpendicular to the axis, pass through each of those points. Then, between any two asymptotes there may be curves of the form given in Fig. 2 of the Introduction. These are represented [here] by DEFI. . ., D′E′F′I′. . ., whose first leg ED would be asymptotic and repulsive, the last leg SV attractive. In each, the interval EN, where the arc of the curve is winding, is exceedingly small compared with the interval near S, where the arc for a very long time continues closely approximating the form of the hyperbola having its ordinates in the inverse ratio of the squares of the distances" (Boscovich 1763, § 171)

In this passage, Child's translation rendered "*mundia*" as "universes", suggesting to some modern commentators that Boscovich was endorsing a strong version of the multiverse. More prosaically, however, he seems to refer to a plurality of size orders within the same universe, which have no way to exchange material. As Boscovich explained, he was referring to worlds "*ut ita dicam*" ("so to speak"), and not to actual, real worlds. Those metaphorical worlds are actually aggregates of different, incomparable sizes (there are *mundi minorum dimensionum*, as well as *majores*), and each of them is formed by masses that, if seen from a greater world, are similar to points. They cannot communicate because of the properties of the asymptotes, which Boscovich briefly resumed in footnote *i* to the preceding § 168 (whereas the extended argument on the asymptotes is exposed at length in Boscovich 1755a, §§ 103–108).

The asymptotes cause each 'world' to remain confined in itself without relation with others. This is a mechanical property (it only involves motions of points or their aggregates), so Boscovich has included this passage in Part II, devoted to the

---

asymptotes, vel inter binaria quotlibet, et rite ordinato, posset exurgere quivis, ut ita dicam, Mundorum numerus . . ., atque id ita, ut quivis ex iis nullum haberet commercium cum quovis alio; cum nimirum nullum punctum posset egredi ex spatio incluso iis binis arcubus, hinc repulsive, et inde attractive; et ut omnes Mundi minorum dimensionum simul sumpti vices agerent unius puncti respectu proxime majoris, qui constaret ex ejusmodi massulis respectu sui tanquam punctualibus, dimensione nimirum omni singulorum, respectu ipsius, et respectu distantiarum, ad quas in illo devenire possint, fere nulla; unde et illud consequi posset, ut quivis ex ejusmodi tanquam Mundis nihil ad sensum perturbaretur a motibus, et viribus Mundi illius majoris, sed dato quovis utcunque magno tempore totus Mundus inferior vires sentiret a quovis puncto materiae extra ipsum posito accedentes, quantum libuerit, ad aequales, et parallelas quae idcirco nihil turbarent respectivum ipsius statum internum."

"application of the theory to mechanics"; however, it has an obvious physical significance, which is explored in the following paragraph, pertaining to Part III:

> If universal gravitation extends indefinitely with the same law and the same asymptotic leg [*crure*], not only our solar system indeed, but the whole corporeal nature, would recede—little by little in truth, but still continuously—from the state in which it was established, and it all would necessarily fall to destruction; all matter would, in time, be conglomerated into one shapeless mass, since the gravity of the fixed stars on one another will not be cancelled by any oblique or curvilinear motion. That this is not the case cannot be absolutely proved; nevertheless, a theory that opens up a way to avoid this universal ruin, as my theory does, seems to be more consonant with the Divine Providence. For it may be that, as I remarked in § 170, the last arc of my curve, which represents gravity, after it has reached distances greater than the greatest distances from the Sun of all the comets that belong to our solar system, begins to depart very much from the hyperbola having as its ordinates the inverse square of the distances; perhaps it will cut the axis again and be twined about it. In this way, the whole aggregate [*totum aggregatum*] of the fixed stars, together with the Sun, might be a single particle of an order higher than those of which the system itself is composed, and it might, in turn, belong to a system immensely greater still. It may even be the case that there are several of such orders of particles, so that particles of the same order are completely separated from one another without any possible means of getting from one to the other along the various asymptotic arcs to my curve, according to what I revealed in § 171. (Boscovich 1763, § 405)

This is certainly a bold conjecture about the (hierarchic) structure of the Universe being composed of a plurality of aggregates of different orders of magnitude. From this perspective, it might even be the case that the Universe contains more "Solar systems" like ours, for, on the basis of § 171, in each of the "worlds" between any two asymptotes, the last "leg" of the curve is nearly proportional to the inverse square of the distances (i.e., a quasi-Newtonian gravity).[32] In this sense, this conjecture is more a physical hypothesis about the distribution of matter in *this* Universe than evidence that Boscovich was a forerunner of the strong idea of a multiverse as "an ensemble of universes characterised by all conceivable combinations of initial conditions and fundamental constants."[33]

[32]Note that, according to Boscovich, the cases of intermediate asymptotes instantiated in Fig. 5.1 may only occur on the largest scales: "When, if ever, this . . . case occurs in our curve, then indeed no point situated on either side of [the limit points where the intermediate asymptote cuts the x-axis] will be able to pass through it to the other side, no matter what the velocity with which it is impelled to approach towards, or recede from, the other distances at which point; for the infinite repulsive area, or the infinite attractive area, will prevent such passage. Now, it can easily be derived from this that this case cannot happen at any rate in the distance lying between the diameters of the smallest particles visible under the microscope and the greatest distances of the stars visible to us through the telescope, for light passes with the greatest freedom through the whole of this interval. Therefore, if there are ever any such asymptotic limit-points, they must be beyond the scope of our senses, either superior to all telescopic stars or inferior to microscopic molecules" (Boscovich 1763, § 188). I will return to the mathematical underpinnings of this in Sect. 6.5.6.

[33]I borrow this phrase from Carter (1974); this is, nowadays, the most popular version of the notion of the multiverse. Of course, there are many weaker versions, ranging from a plurality of systems (possibly infinite in number), each of them gravitating around its particular sun, to multiple causally disconnected space-time portions originating from the Big Bang. For a historical account of the subject, see Kragh (2009). On any account, Boscovich's physical Universe as described in the *Theoria*—the unique Universe we inhabit—was a static one, composed of homogeneous points of

In fact, the background of his conjecture is the so-called cosmological paradox that Bentley proposed to Newton: even with the immense distance between the stars and their uniform distribution, gravity is always attractive, although extremely weak. This would necessarily bring the Universe to collapse in upon itself, perhaps in an enormously large, but finite, amount of time. As Boscovich stated, "all matter would, in time, be conglomerated into one shapeless mass." Newton asked this question in the *Opticks* (Query 28): "What hinders the fix'd Stars from falling upon one another?" (in the Latin version Boscovich was acquainted with: "*Et Quidnam est quod impedit, quominus Sol & Stellae fixae in se mutuo irruant?*"). One answer could be that the stability of the stars proves God's providence in maintaining the Universe in equilibrium. (This was probably Bentley's answer, to which Newton might have also been sympathetic).[34] However, for reasons involving the Jesuits' traditional natural theology, already, in *De viribus vivis* (§ 59), Boscovich attempted to arrive at a different solution.[35] The curve of forces prescribes that, at some distances, points are determined to recede from one another or to approach one another, but it does not fix those distances. It is arguable, indeed, that at the largest, interstellar scale, the force becomes repulsive again, so that the fixed stars would preserve their relative position and a paradox does not occur. Thus, the Universe would obey a sort of cohesion law that is iterated on increasingly large scales: "The difficulty, which has been repeatedly brought against the Newtonian theory on account of this necessary mutual approach of the fixed stars, disappears altogether in my theory. At the same time, we have now passed on from gravity to cohesion" (Boscovich 1763, § 406).

---

matter (which, in turn, cause inhomogeneous aggregates to be formed), and homogeneous with respect to its properties.

[34]On the cosmological paradox, Newton's response to Bentley and other attempted solutions in Boscovich's time, see Guicciardini (1996).

[35]"What if ... even the fixed stars were to be placed on some limits of all the attractions and repulsions? That is, suppose that the curve ... were also to disagree at the biggest distances (above the planets) ... with the hyperbola expressing gravity, which decreases according to the law of the inverse square of the distances. It would cut the axis again, perhaps at many other points. Would not the fixed stars have been preserved at nearly the same distance from one another, without falling upon one another [*nec in se mutuo irruerent*], and would not the whole world be structured like one of those larger particles?" (Boscovich 1745, § 59). The passage "*An non distantiam servarem a se invicem proxime eandem, nec in se mutuo irruerent*" ("Would not the fixed stars have been preserved at nearly the same distance from one another, without falling upon one another") echoes Newton's Latin text in the above-quoted *Optice*: "*Et Quidnam est quod impedit, quominus Sol & Stellae fixae in se mutuo irruant?*" For this reason, I have rendered the verb *irruo* in the same manner as in the *Opticks*, "to fall upon". To be sure, Boscovich had precise knowledge of the Latin edition of the *Opticks*, which is the most quoted work in Boscovich (1757) and is often quoted in the *Theoria* as well. Also note that the *Optice* (in the Paduan edition of 1749, also containing Newton's *Lectiones Opticae*) is listed among the works that he had in his own possession in a document *De libris qui desiderantur pro classe Mathematica Universitatis Ticinensis* (1766 ca.), a list of works that he considered necessary for a library of mathematics to be constituted at the University of Pavia. The document is preserved at the Archivio di Stato di Milano (*Autografi*, cart. 115). See also Guzzardi (2015, 351, 2016, 13, n. 27).

On this background, the above-quoted § 171 does not expose a theory of many worlds or a multiverse, but it does convey the no less original idea of a unique Universe, possibly composed of a plurality of "Solar systems" and viewed as a potentially never-ending series of aggregates of points, which, at gradually larger scales and considering reciprocally incommensurable orders of magnitudes, tend to display a globally uniform behavior, although the curve may change locally.

However, the *Theoria* also briefly considers the possibility of radically different "worlds", formed by points that differ in species. As a premise, Boscovich (1763, § 517) suggests that matter might not be as homogeneous as he has supposed so far. Then, if there were to be in nature phenomena "that cannot be explained by a single kind of matter [*materiae genus*], we shall make use of many different kinds of points [*genera punctorum*] with many laws that differ from one another," which can be combined further. Of course, the laws can be expressed by curves, and it may happen that some of those curves "would have something in common, such as the asymptotic arc of impenetrability, or the arc of gravitation; while some might be considerably different from others." Now, since such laws govern the aggregation of points into definite compounds, "certain kinds and certain differences will be obtained [*ut habeantur quaedam genera et quaedam differentia*], which would distribute the elements of the bodies in certain classes." This opens an interesting potential scenario, sketched in the subsequent paragraph: "In some of these classes, a null force could be admitted, then the substance of one of these kinds will flow absolutely freely thorough a substance of another kind without any collision . . . and thus the substance would pass through with real impenetrability and apparent compenetration" (Boscovich 1763, § 518). Note that impenetrability is *real* because it is governed by a determinate force law that only applies to points of the same species, whereas the compenetration is *apparent* because it is—so to speak—the observed phenomenon.

It is not easy to explain why Boscovich mentioned these possibilities. Baldini (1992a, 25) has suggested that his Jesuit education led him to extend "the area of the mental experiments" to include the space of the logical possibilities and view them as feasible in principle. On the other hand, a Jesuit author as influential as Francisco Suárez tended to use his own theory of "imaginary space" as "*ens rationis cum fundamento in re*" in order to explain how different substances—the body, the soul, or God Himself—can be co-present in the same space at the same time (i.e., they can occupy an identical point of space-time) while remaining different substances (without compenetration, as Boscovich would put it).[36] This agrees with Boscovich's further conjecture that "there might be a large number of material and sensible worlds existing in the same space, separated one from the other so that one would have no communication, and one could never acquire any notion of the other" (Boscovich 1763, § 518). This is certainly a "speculation concerning multiple worlds," as Kragh (2009, 534) put it; it is left to the reader to judge whether, and

[36]For an introduction to Suárez's doctrine of *entia rationis*, or beings of reason, see Shields (2012).

to which extent, such a speculation also agrees with (some of the multiple versions of) the modern idea of the multiverse.

4.4 The Number of Points of a Body: Boscovich's Notion of Mass and Its Source

The aggregational view sketched in the previous section, along with Boscovich's concept of the material point, brings about a remarkable consequence: the mass of a body is an emergent physical property of that body as an aggregate of points.

As we have seen, we can consider bodies as a compound of points-at-a-distance (cohesion of points is an effect of their mutual distances). Because distances separating points must necessarily be void of matter—given that the law of forces forbids immediate contact—bodies can be viewed as a compound of matter (material points) and vacuum. In Boscovich's own words, "The primary elements of matter ... are scattered [*dispersa sunt*] in an immense vacuum ... I do not consider that vacuum is strewn [*disseminatum*] within matter, but that matter is strewn within vacuum and floats in it" (Boscovich 1763, § 7). This helps us to understand Boscovich's notion of the mass of a body: this is defined, in Newtonian terms, as the quantity of matter; however, since vacuum does not compose matter, quantity of matter only depends on the quantity of material points—in other words, "the mass of a body is the total quantity of matter pertaining to [a] body, which will be to me precisely the same thing as the number of points that go to form that body" (Boscovich 1763, § 378).[37]

Yet, this notion of mass is puzzling in one respect at least. Boscovich's intention in this paragraph can hardly have been to provide a feasible measure of the mass of a body. (Otherwise, how should we count points?) Moreover, there was no need to do so, since such a measure could be extrapolated from Newton's explication of Definition I of the *Principia*: mass "can always be known from a body's weight,

[37]Note that this view had already emerged in the 1748-drafted *De materiae divisibilitate*—that is, presumably at the time of the Camaldolese *Ur-Theorie*—, in which Boscovich wrote that "matter must necessarily be homogeneous, and gravity itself must be ... directly proportional to the number of points, that is, to mass [*numeri punctorum seu massae*]" (Boscovich 1757, § 46). However, *mass* is a somewhat ambiguous term in Boscovich's works. It sometimes stands for "body," but also has a wider meaning and may apply to any aggregate of points that can be regarded as a whole because of the interplay of attraction and repulsion that causes a state of cohesion (see, e.g., Boscovich 1755a, §§ 114–119). It is worthwhile noting that Boscovich's notion of a "material point" is very different from the apparently similar notion of a mass point, which emerged in Western science over a long span of time (see Foligno 2018). Of course, Newton played a fundamental role in elaborating a "gravitational" idea of a material point by conceiving the center of gravity of a body as the geometric point of space in which mass, or weight, is thought to be concentrated: see Truesdell (1968, esp. 92–93, 106–107) and Guicciardini (1999, 68–71, 2009, 251, 2011, 165–166).

for—by making very accurate experiments with pendulums—I have found it to be proportional to the weight" (Newton 1687, 404). It can also be extrapolated from many other places. And by any account, Boscovich (1763, § 381) seemed to acknowledge Newton's use in stating that "whatever is usually said about comparisons between mass, volume and density, everything is in agreement with what I say. Mass is, so to speak, the product of volume and density"—which is clearly reminiscent, again, of Definition I of the *Principia*: "Quantity of matter is a measure of matter that arises from its density and volume jointly ... I mean this quantity whenever I use the term 'body' or 'mass' in the following pages" (Newton 1687, 403–404).

However, he might have pondered another reason to introduce his own idea of mass as the number of points that form a certain body. A similar view had been expressed some years earlier by Leonhard Euler in his *Mechanics* (1736), a work of which Boscovich had been aware since 1740, but with which he asserted his disagreement in fundamental ways.[38] In the preface, Euler (1736, 9) introduced the idea of operating on "infinitely small bodies" that can be considered as points. This should have suggested to him "the division of this work, so that, in the first place, I will investigate the motion of infinitely small bodies and, as it were, points [*corporum infinite parvorum et quasi punctorum motum*], but then I will go forth to bodies of finite magnitude, either rigid or flexible or made up of parts completely disunited one from each other [fluids, in modern terms]."

However, the nature of such infinitely small bodies remains elusive in Euler's account. Are they merely a useful mathematical fiction? Or are they real, discrete particles of which extended bodies are composed? The "mature" Euler would have possibly excluded the latter interpretation, but elsewhere, his inclinations appear somewhat ambiguous (Stan 2017). For instance, in the "Scholium General" to Chap. 1 of the *Mechanics* ("On motion in general"), he advanced that the physical laws that a free body in inertial motion obeys "properly pertain to infinitely small bodies, which can be considered as points,"[39] with no implications about the

[38]Let us remember that, according to Boscovich (1740), inertia "can be demonstrated neither from the phenomena nor through metaphysical arguments": see above, Sect. 3.1. In particular, "nothing shall ever be achieved from metaphysical arguments if the nature of bodies is not known. And this only unravels to us through the phenomena. For this reason, we seem to see a vain effort in that of the very learned Leonhard Euler, who endeavors to demonstrate [*nititur demonstrare*], in the very beginning of his *Mechanics* ..., such inertia of the bodies, and yet without succeeding" (ibid., III). On the contrary, Euler (1736, 31, § 75) stated that "although we demonstrated both the persistence in the state of rest and the uniform, rectilinear prosecution of motion from the principle of sufficient reason [ibid., 27, § 56 and 29, § 65], we have already noticed that this is not the efficient cause of the phenomena, but it is placed in the nature of bodies themselves. Thus, the cause for which they preserve their state depends on the nature of bodies, and [this cause] is what is called *vis inertiae*."

[39]"*Istae motus leges, quas corpus sibi relictum vel quietem vel motum continuando observat, spectant proprie ad corpora infinite parva, quae ut puncta possunt considerari*" (Euler 1736, 38, § 98). Euler also stated the following: "For in bodies of finite magnitudes, of which the individual parts have their own motions, any part will endeavor to observe these laws, which, however, will not always be possible because of the state of the body. The body therefore will

constitution of matter. However, in Chap. 2, when dealing with "the effect of forces on free-acting points," Euler's conception of matter seems to surface.

Proposition 16 expresses a theorem according to which "a force q acting on a point B has the same effect as a force p acting on a point A, if taken for granted that q: $p = B$: A" (Euler 1736, 50: here and in the following pages, I will use capital letters for points and small letters for forces, whereas in Euler's text, only small letters are used). This implies that the "greater" the entities to which the force is applied, the greater the force will have to be in order to have the same effect. This is cleared by the following corollary (ibid., 50): "In order that a greater point be induced to the same velocity as a smaller one, a greater force is needed, and that force ought to be the same degree greater as the former point is greater than the latter. [*Ad eandem ergo maiori puncto celeritatem inducendam quam minori, opus est maiori potentia, idque tanto maiori, quanto illud punctum maius est quam hoc*.]"

But how should we interpret expressions such as "*maior punctum*" and "*minor punctum*", respectively meaning a greater point and a smaller point? Size cannot be involved, because points are zero-dimensional objects. Thus, it seems perfectly reasonable to interpret this as a difference in *masses*. Therefore, mostly based on this passage from the *Mechanics*, Euler's scholars have ascribed to him the introduction of "the precise concept of *mass-point*."[40] Yet, this is only partly true. I agree that Euler's work represents a crucial step in the formation of the modern idea of a mass point (or material point) as a mass that does not have any real value but is associated with null extension. On the other hand, as I have argued, Newton represents another of such crucial steps, although in a different direction, which I termed gravitational (see above, footnote 37) and that cannot be revealed further here. To make matters more complicated, Euler slightly changed his definition of infinitesimal small bodies in the years that followed, in a manner that brought him very near to our modern idea of a material point.[41]

---

follow a motion that is composed of the endeavors of the individual parts ... The diversity of bodies therefore will supply the primary division of our work. In the first place, we will consider infinitely small bodies, i.e., those that can be considered as points [*contemplabimur corpora infinite parva seu quae tanquam puncta spectari possunt*]. Then, we will approach bodies of finite magnitudes that are rigid and do not undertake to shape changes. Thirdly, we will consider flexible bodies. Fourthly, we will deal with bodies that admit extension and contraction. Fifthly, we shall subject to examination the motions of several separated bodies, some of which hinder each other from executing their motions as they attempt them. Sixthly, at last, the motion of fluids will have to be treated. ... I begin with the motion of free points perturbed by powers of any kind [*Incipio igitur a motu punctorum liberorum a potentiis quibuscunque sollicitatorum*], because these, left to themselves, will follow their motion, as we have already shown in this chapter."

[40] I use here a phrase from Truesdell's influential *Essays in the History of Mechanics* (1968, 107, italics in the original text) as a label for that which has been presented as a historiographical common place. See, e.g., van der Waerden (1983, 276), Gaukroger (1982), and Stan (2017).

[41] As noted by van der Waerden (1983, 272), in Euler's *Decouverte d'un nouveau principe de mecanique* (1756), an infinitesimal small body ("*corps infiniment petit*") is characterized as a body "*ou dont toute la masse soit reunie dans un seul point*."

However, I argue that, in Proposition 16, Euler did not present mass (as distinct from inertia) as a primitive concept; rather, he *constructed* the concept of mass starting from *his* primitive concepts: infinitely small bodies *sive* points. As in the case of Boscovich, these are defined in kinematic-dynamical terms only: they are entities that tend to perform a motion when subject to a "force" (*potentia*, as Euler called it). Let us consider the demonstration of the statement referred to above:

> Let be $q = \boldsymbol{n}p$; it will be $B = \boldsymbol{n}A$. Now, let point $\boldsymbol{n}A$ be conceived as divided into $\boldsymbol{n}$ equal parts, each of them be $= A$. Let each part be disturbed by the $\boldsymbol{n}$-th part of the force $\boldsymbol{n}p$, i.e., by a force. As a consequence, any part whatever will be pulled by its force in the same manner in which the same point A is pulled by the force p . . . It is also clear that the following two cases are reduced to the same case and do not differ from each other: either that the point $\boldsymbol{n}$A is pulled by the force $\boldsymbol{n}p$, or that any part a whatever of the point $\boldsymbol{n}$A is pulled by a similar part p of the force $\boldsymbol{n}p$, insofar as the parts are not separated from each other. For this reason, the thesis is certain that $\boldsymbol{n}A$ is impelled by $\boldsymbol{n}p$ in equal manner as the point A by the force p.[42]

Euler aimed to exploit the parallel between infinitesimal small bodies and points in all its potential. A geometrical point *cannot* obviously be conceived as being partitioned in any number of parts, no matter whether they are equal or not, for the good reason that "a point is that which has no part," as reads the first definition of Euclid's *Elements*. Euler's points, however, may indeed have parts, since they are *points-as-it-were*; they *can be considered* as points, but they are actually infinitely small *bodies*.

Euler suggested taking one of such points, such as A, as a basic unit of measure and conceiving of any other point B as a multiple of A: in fact, as an aggregate of n basic units. Now, it is perfectly reasonable to speak of greater and smaller points, namely, infinitely small bodies composed of a greater or smaller n-multiple of A-points, without commitment regarding the "mass" of an individual point. Conversely, the notion of mass depends on this. In the subsequent scholium, he expressed the following:

> This proposition [16] encompasses the grounding principle for measuring the *vis inertiae* [*fundamentum complectitur ad vim inertiae metiendam*], for here relies all reason whereby in mechanics the matter of bodies, nay the mass [*corporum materia seu massa*], must be considered. One should carefully take into consideration the number of points of which the body to be moved is composed and to which the mass of the body ought to be posited as proportional. Points, indeed, should be thought of as equal to each other, not in the sense that they are equally small, but in so far as the same force raises equal effects. If we conceive of all matter as being divided into equal points or elements of this kind, we necessarily will

[42] "*Ponatur* q = np, *erit* b = na. *Concipiatur iam punctum* na *in* n *partes aequales divisum, quarum quaelibet erit* = a; *harum partium unaquaeque sollicitata sit a parte n-sima ipsius potentiae* np, *id est a potentia* p. *His positis quaevis pars eodem modo trahetur a sua potentia, quo punctum ipsum* a *a potentia* p. *Neque vero hae puncti* na *partes a suis potentiis sollicitatae a se invicem segregabuntur; sed perpetuo unitae manebunt, si quidem initio fuerint coniunctae. Perspicuum autem est hos duos casus eodem redire nec a se invicem discrepare, sive punctum* na *a potentia* np *trahatur, sive quaevis puncti* na *pars* a *a simili parte* p *potentiae* np *trahatur, dummodo partes non a se invicem divellantur. Quapropter constat propositum* na *aeque a potentia* np *urgeri ac punctum* a *a potentia* p" (Euler 1736, 51). In my translation, I used capital letters for points and small letters for forces. I chose to write $\boldsymbol{n}$ (the number of "parts") as a small, italicized bold character in order to distinguish it from points, as well as from forces.

determine the quantity of matter of whatever body through the number of points that it is composed of [*Si igitur universam materiam in huiusmodi aequalia puncta seu elementa concipiamus divisam, quantitatem materiae cuiusque corporis ex numero punctorum, ex quibus est compositum, aestimari necesse est*]. (Euler 1736, 51)[43]

Let us reframe and comment on the various steps of Euler's discourse:

1. We assume that the theorem of Proposition 16, If q: $p = B$: A, is demonstrated; then, force q acting on point B has the same effect as a force p acting on point A.
2. Let us conceive of q as a force $\boldsymbol{n}$ times greater than p; correspondingly, with A and B being infinitesimal small bodies that we can consider as points, let us conceive of B as a point $\boldsymbol{n}$ times greater than A. Then, $\boldsymbol{n}A$ will need force $\boldsymbol{np}$ in order to obtain the same effect of force p exerted upon A.
3. For that reason, passages 1–2 (above) indicate how inertia is to be measured, for *vis inertiae* has been defined as the capacity to resist changes in the state of motion (Euler 1736, 31–32: Def. 9), and $\boldsymbol{n}A$ will exert $\boldsymbol{n}$ times the resistance of A.
4. Passages 1–2 also suggest a clear image of the meaning of "greater than": One thing is (proportionally) greater than another when the same force raises a (proportionally) smaller effect, since that thing exerts a (proportionally) greater resistance. Thus, things are "equally small" when "the same force raises equal effects."[44]
5. Now, let us assume that "the matter of bodies *seu* mass" is composed of equally small points. This means that the mass of a certain body ought to be posited as being proportional to the number of points composing it. In other words, we can determine the mass of a compound body by "counting" its composing points. (A corollary is that it is meaningless to speak of the mass of an individual point—and note that Euler avoids doing so—because the individual point is the unit of mass.)

Notwithstanding his reservations about the status of the principle of inertia in Euler's *Mechanics* (as expressed in Boscovich 1740, III), the concept of mass that Boscovich would develop in his natural philosophy is reminiscent of Euler's view,

[43]In particular, the demonstration that "the force of inertia of a body whatever is proportional to the quantity of matter of which it is composed" is given in Proposition 17 and is essentially based on Euler's Definition 9, where inertia is described as the tendency to resist changes in the state of motion (see Euler 1736, Def. 9, §§ 74–76, 31–32), and Proposition 16, showing that, if force p is needed in order to move point A, then force $\boldsymbol{np}$ is needed in order to move an infinitely small body B composed of $\boldsymbol{n}$ points as A, with each of them endowed with the same inertia. But, of course, all of this holds under the Newtonian premise that—as Definition III of the *Principia* reads—"inherent force of matter is the power of resisting by which every body, as far as it is able, perseveres in its state either of resting or of moving uniformly straight forward." Moreover, as Newton clarifies in the subsequent explication, "this force is always proportional to the body and does not differ in any way from the inertia of the mass except in the manner in which it is conceived. Because of the inertia of matter, every body is only with difficulty put out of its state either of resting or of moving. Consequently, inherent force may also be called by the very significant name of force of inertia" (Newton 1687, 404).

[44]As remarked by Lützen (2005, 23–24, 135–136), this would lead to Ernst Mach's (and others') operational definition of mass, as well as to Hertz's concept of *Massentheilchen* in his *Mechanik in neuem Zusammenhange dargestellt* (1894).

even if in a more realistic vein. As we have seen, material points are real points of matter, obtained through a process of division. But, like Euler's "infinitely small bodies, as if it were points," they are described only in terms of inertia and attractive-repulsive forces according to distances. And most of all, like Euler's points, Boscovich's material points involve the foundation of the notion of mass.

4.5 A Three-Layered Metaphysics of Space

A final fruit of Boscovich's investigation of the primary elements of matter and the aggregates that they can form was a new analysis of the "real" space in which bodies are found and physical actions take place.

He probably soon realized that his material points involved a sort of feedback effect on the Aristotelian conception of space from which he started, in so far as they result from a physical-metaphysical procedure based either on Aristotelian doctrines or on the idea that points, as far as they are material, are endowed with inertia and attractive-repulsive forces. This turn was fully accomplished in a couple of supplements to the first volume of Stay's (1755) Newtonian poem, which were ultimately included, virtually without changes, in the *Theoria*.[45] However, the relationship between the finite divisibility of matter and the continuous character of geometric space must have preoccupied him as early as 1748. Some connections between the notes and supplements to Stay (1755) and the 1748-drafted *De materiae divisibilitate* are evident. For instance, in a footnote to Book I of Stay's poem, he described space and time as "the most difficult among the metaphysical problems" (Stay 1755, 25: footnote *a* to Book I, v. 607). Referring to Augustine of Hippo, he stated, "*Si me interrogas nescio, si non interrogas, scio*" (the correct citation, from Augustinus, *Confessiones*, XIII, XI, 14, reads: "*Si nemo ex me quaerat, scio; si quaerenti explicare velim, nescio*"). Correspondingly, in *De materiae divisibilitate*, he had stated that the "constitution [*constitutio*] of space and time ... pertains to metaphysics." Thereafter, he commented, "if we are not asked about them, we know what they are; but if somebody asks, we do not know anymore" (Boscovich 1757, § 13). Even some of the 1748 correspondence suggests that the collaboration for Stay's Newtonian poem traces back to the same period in which he worked on his response concerning the divisibility of matter, which originated in both *De materiae divisibilitate* and the Camaldolese *Ur-Theorie*.[46]

[45]Entitled "Of space and time" and "Of space and time, as we know them," they were originally published as supplements VI and VII to the first volume of Stay (1755). They were reprinted, virtually without changes, in the *Theoria*. First, they formed Supplements III and IV to the 1758 edition. Thereafter, they became Supplements I and II in the 1763 edition.

[46]On March 16, 1748, he wrote the following to his brother Bartolomeo: "Every evening, then, I dictate his Physics to father Beno [i.e., Benedict Stay] for one or two hours," hinting at Stay (1755). On Boscovich's role in Stay's compositions, see Tacconi (1994a, b).

Again, it is in the 1748-drafted text of *De materiae divisibilitate* that Boscovich introduced one of the key ideas that he would explore further in the supplements on space and time. The space with which geometry deals is "completely imaginary and not real, something existing without bodies and motions" (ibid.; see also Boscovich 1757, § 57). This is—as he put it in the 1755 supplements—an "imaginary space," as opposed to the "real space" (with the same applying to time):

> For each point of matter ... I admit two real kinds of modes of existence, of which some pertain to space and others to time; and these will be called local and temporal modes, respectively. Any point has a real mode of existence, through which it is where it is; and another, due to which it exists at the time when it does exist. These real modes of existence are to me real time and space; the possibility of these modes, hazily apprehended by us, is, to my mind, empty space and again empty time, so to speak; in other words, imaginary space and imaginary time.[47] (Boscovich 1763, S.I, § 4; in the following, I will use the abbreviation S. followed by a Roman numeral and, when needed, by the relevant paragraph, to refer to the "Supplements" of the *Theoria*.)

Expressions like "real time" and "real space" do not occur in *De materiae divisibilitate*, in which material points are distinguished from mathematical points, as long as the former are "real, can be endowed with real motion and have real properties" (Boscovich 1757, § 13; see also ibid., §§ 17–18, 26). However, since material points necessarily remain at some definite distance (for repulsion infinitely dominates at infinitely small distances), they cannot "compenetrate". In other words, as I mentioned earlier, "different [material] points necessarily occupy different points of space" (ibid., § 26).

This suggests a threefold distinction between material points, the points of space that they actually occupy—i.e., their real coordinates—and space in a purely geometrical sense, as a set of all possible coordinates. It is exactly this three-layered structure that Boscovich developed and articulated in the supplements.[48] For the sake of simplicity, and because of its relevance to the concept of the material point, let us consider space alone and proceed from possibility to reality (although it is worth remarking that Boscovich, in consonance with the Jesuit philosophical tradition, considered the constitution of space and that of time, *via* their common continuous structure, as perfectly analogous: see Baldini 1992a, 36–37). Any particular point has an infinite *possibility* of positioning. In other words, we can insert a point of space wherever we want. Moreover, as observed above, geometric continuity, the infinite divisibility of space, and the endless insertibility of boundaries

[47] Pro singulis materise punctis ... admitto bina realia modorum existendi genera, quorum alii ad locum pertineant, alii ad tempus, et illi locales, hi dicantur temporarii. Quodlibet punctum habet modum realem existendi, per quem est ibi, ubi est, et alium, per quem est turn, cum est. Hi reaies existendi modi sunt mihi reale tempus, et spatium: horum possibilitas a nobis indefinite cognita est mihi spatium vacuum, et tempus itidem, ut ita dicam, vacuum, sive etiam spatium imaginarium, et tempus imaginarium (Boscovich 1763, S I, § 4).

[48] This is also recognized in a footnote added in 1757 to the original 1748 draft of *De materiae divisibilitate*: "From that theory, we can positively deduce that there is an interpolated existence that is not a pure nothing, since it is constituted by points' real modes of existence, which, in turn, constitute a real relation between distances, as I explain in more detail in the dissertation *De continuitatis lege* [Boscovich 1754b] and in the *Supplements* to Stay's work [Boscovich 1755b]."

(points) are essentially the same. As Boscovich explained, on the one hand, "beyond and between two real points of position of any sort, there are other real points of position *possible*"; on the other hand, "there will be a real divisib*ility* to an infinite extent of the interval between two points, or, if I may call it so, an endless insertib*ility* [*interseribilitatis*] of real points" (Boscovich 1763, S I, § 8; emphasis added). All of this is merely possible: We *can* do something that we have not already *done*. In other words, this is an imaginary space, originating from a sort of abstraction practice:

> In this way, so long as we conceive of as possibles these points of position [*loci puncta*], we have infinity of space and continuity together with infinite divisibility. With existing things, there is always a definite limit, a definite number of points, a definite number of intervals; with possibles, there is none that is finite. The abstract concept of possibles, excluding as it does a limit due to a possible increase of the interval, a decrease or a gap, gives us the infinity of an imaginary line and continuity; such a line has not actually any existing parts, but only possible ones. Also, since this possibility is eternal and necessary, since it was true from eternity and of necessity that those points might exist in conjunction with those modes, such imaginary, continuous, and infinite space is eternal and necessary as well. However, it is not something existing, but something that is merely potentially existing, and indefinitely conceived by us [*non est aliquid existens, sed aliquid tantummodo potens existere, et a nobis indefinite conceptum*]. The immobility of this space will come from immobility of its individual points. (ibid., S I, § 9)

Some of these merely possible points, however, are brought to actuality by becoming associated with some material points. (Incidentally, it is this association that we call spatial and temporal coordinates.)[49] Every point of matter, insofar as it exists and can be described in terms of its kinematic state, can be expressed through its "points of position", causing it to become real. These are, as Boscovich described them, *real* space-time modes of existence: the actualized representatives of possible points in geometric space and imaginary time.

The fact that they are actualized possibilities in the geometric space has two important consequences. First, they are "immovable and unvarying in their order" ("*immobiles, ac in suo ordine immutabiles*": ibid., § 5), for they have only changed their status from possible to real, but their relative positions are unchanged. In other words, as we have mentioned above, Boscovich maintained that motion pertains to material points, but not to imaginary, geometric space (see also Boscovich 1757, § 57). In the second place, if only material points move, the real points are "produced" when a moving material point occupies and actualizes them for an instant, but they also "perish" ("*Modi illi reales singuli & oriuntur, ac pereunt*": Boscovich 1763, S I, § 5) as soon as the actualizing material point has left one geometric point for a new one, calling the latter to real life and throwing back the former into the realm of mere possibilities. We could visualize this with a modern image by representing the geometric space (imaginary in Boscovich's sense) as an infinite plane formed by

[49]They [i.e., the space-temporal modes of existence] afford the foundation of a real relation of distance, which is either a local relation between two points or a temporal relation between two events. Nor is the fact that those two points of matter have that determinated distance anything essentially different from the fact that they have those determinated modes of existence, which necessarily alter when they change the distance (Boscovich 1763, S I, § 5).

imaginary inactive leds arranged very thickly, each led representing a geometric point. In order to make our image more satisfying and fitting to Boscovich's conception, let us imagine that leds become visible only when lit, and that they are continuous, i.e., we could interpolate any two leds with any number of other leds. Each led can be activated by a moving, real material point, which turns it on solely for the instant in which the material point exactly occupies this particular led. The led is now—and only now—lit and visible. Or, in Boscovich's terms, a geometric point has been brought to actuality, becoming real (in our image, *being real* is equal to *being visible* or *being lit*): a real point has been *produced* (or *turned on* in our image). However, as soon as the material point moves away, the led turns off and becomes inactive and invisible again. In Boscovich's terms, the real point has *perished* and recovered its status of possibility in the imaginary space.

Now, since extended matter is formed by material points-at-distance (i.e., matter is discretely constituted), in any instant of time, not all of our leds will be active, hence lit and visible, but only those occupied by a material point. In other words, whereas the set of possible points (inactive, invisible leds) is infinite *ex hypothesi*, both in the case of the material and the real points of space (lit, visible leds), we are only faced with finite sets. Thus, it is not only matter that is finitely divisible; real space also cannot be divided to infinity, simply because it is formed by the points that are really associated with the finite set of material points: "Since existent points of matter always have some distance between them, and are finite in number, the number of local coexisting modes is also always finite [*finitus est semper etiam localium modorum coexisteintium numerus*]; and from this finite number, we cannot form any sort of real continuum" (Boscovich 1763, § 142).[50]

[50]Child's translation incorrectly reads as follows: "The number of local modes of existence is also always finite" (in fact, the whole passage is ill-translated). Co-existence is the key-feature here, for Boscovich emphasizes that, *at any instant of time*, we are faced with a finite set of real points, but, of course, real points are potentially infinite in number as time passes. So, Boscovich immediately adds: "But indeed, to me [*mihi*] the imaginary space is the possibility, confusedly cognized, of all local modes [*possibilitas omnium modorum localium confuse cognita*], which we conceive simultaneously by precisive cognition [*simul per cognitionem praecisivam concipimus*], although they cannot all exist simultaneously [*licet simul omnes existere non possint* . . .]. So long as we keep the mind free from actual existence and, in a series of possibles consisting of an infinite number of finite terms [*in possibilium serie finitis in infinitum constante terminis*], we mentally exclude the limit both of least and greatest distance, we form the idea of continuity and infinity in space." The expression *cognitionem praecisivam* implicitly refers to Francisco Suárez's doctrine of the formation of the universals. According to him, they "can come up either by the absolute precisive act of the passive [or potential] intellect, by which nature is grasped and separated from its individuality according to its essence (nature) and its precise formal '*ratio*', or they can be produced by the collative or comparative act, by which nature, directly prescinded from particulars, is related to things, in which it extramentally exists, and from which it has been abstracted" (Heider 2015, 184). A further reminiscence of the Suárezian doctrines is the expression *confuse cognita*, meaning things that are cognized "as a whole" and not distinctly. On Suárez, his (and others') "epistemology of universals", the idea of "confused concepts", and "the reality of the possibles", see Courtine (1990, 293–321), Doyle (2010, 21–39, 46–49), Heider (2014, esp. Chap. 2, 2015). Again, Boscovich's concepts of "real modes of existence" and "imaginary space" are clearly indebted to Suárez (and mediated by Giovanni Battista Tolomei): see Baldini (1992a, 18–19, 35–40). Therefore, Boscovich's case adds interesting and—with the exception of Baldini—so far undervalued evidence to Suárez's influence upon modern epistemology (for an overview, see Sgarbi 2010).

Chapter 5
The Other Labyrinth

5.1 Strategy Changes

In the last chapter, I argued that, via transduction, Boscovich's force curve causes matter (and, indirectly, real space) to be discontinuous. Let us assume that, as it happens with Boscovich's curve, repulsion grows asymptotically at infinitesimal distances; then, let us imagine dividing bodies and their particles through and through. According to the analogy of nature, no matter how we decide to divide particles, this process preserves their inertia, as well as their determinations to approach one another or recede from one another depending on distances. Because of this, the constituent parts of any extended object must be at some definite distance in order that the object is found in a (metastable) state of equilibrium, thus producing cohesion. Thus, bodies are composed of material points-at-a-distance—of matter and void.

Of course, the discrete microscopic structure of matter could be avoided if one assumes the thesis opposite to that in Boscovich's theory, i.e., if the assumption is made that attraction grows as distances diminish. But this was not an option for Boscovich, since it implies—*contra* what he discussed as early as in *De viribus vivis* (Boscovich 1745, §§ 46–47; see the Introduction)—that continuous processes could go broken in nature, for example, in body collisions. For, if there were to be no force repelling bodies at infinitesimal distances, when they approach more and more, their velocities "would totally be extinguished in an instant of time" at the very moment of impact. In this sense, the discrete structure of matter, which I have extensively presented in the previous chapter, is a consequence of the tenet of continuity in natural processes. This chapter aims to explore such a crucial concept.

In the 1745 dissertation on living forces, continuity was expressed in a standard late Aristotelian form, "In nature, nothing happens by jump," and was explained as follows: "Everything that is increased or diminished is continuously increased or diminished, so that it passes from a magnitude to another, by a motion that is always

L. Guzzardi, *Ruggiero Boscovich's Theory of Natural Philosophy*, Science Networks. Historical Studies 60, https://doi.org/10.1007/978-3-030-52093-9_5

continuous, through all the intermediate magnitudes" (ibid., § 45). Two different and almost opposed justifications were provided: first, Boscovich appealed to the circumstance that "no experiment proves that this principle does not hold; lots of them . . . plainly bring us to it," which seems to imply a reference to induction; then, after a couple of paragraphs, he observed that "the principle expressed above is deduced from the analogy and simplicity of nature," which explicitly refers to a process of deduction.[1] Of course, the former method does not exclude the latter; but their mere juxtaposition, without further explanation, signals a certain levity in treating continuity.

The issue of the analogy of nature is particularly critical. As I argued above (see Sect. 4.3), Boscovich seemed to be willing to expand its scope in a sense that Newton would hardly have endorsed. The latter used this principle in order to justify the transduction of observable, macroscopic properties to the microscopic components of matter, but was very cautious about applying it to other realms (e.g., to the optical phenomena). Boscovich, however, considered the analogy of nature as a more general tool for justifying other principles or rules. For example, in his footnotes to Stay's Newtonian poem, he stated that the first and second rules of philosophizing (but not the third) resulted from the ubiquitous analogy and simplicity of nature (Boscovich 1755b, 46 footnote *b*). But appealing to analogy for the purpose of "deducing" the tenet that natural processes do not happen "by jump", as he claimed in 1745s *De viribus vivis*, was by far a bolder move. First of all, how could he think himself to have *deduced* the continuity of processes—physical quantities pass through all intermediates magnitudes whenever they pass from an assigned magnitude to another—from a statement that only endorses uniformity of natural behavior, whatever it means? After all, things could happen in a *uniformly discontinuous* manner as well. Second, how could he consider continuity to have been proven by induction solely because no experiment so far had disproved it—without justifying any legitimacy of the method of induction? It was, frankly, too much, or perhaps too little, for a concept that was supposed to be crucial.

The issue remained unsolved for nearly a decade, until it provided the subject of the long treatise *De continuitatis lege* (Boscovich 1754b), which recognized in it the very foundation of the whole theory of forces. It is not clear whether Boscovich ever realized that the double strategy that he adopted in 1745 in arguing for it—induction plus the analogy of nature—was not entirely satisfying. On the one hand, he probably only later began to think of continuity as something that was more than a tenet shared by most natural philosophers and in terms of a fundamental principle that needed a wide, analytical justification. It is certainly meaningful that, in *De*

[1]The original text of the two quotations reads as follows: "*Communis iam est multorum sententia, nihil in natura per saltum fieri . . . Quidquid augetur, aut minuitur, ita continuo augeri, aut minui, ut ab una quantitate ad aliam motu semper continuo per omnes intermedias quantitates transeatur. Hujus principii nulla experimenta nullitatem evincent: plurima . . . eo nos manifeste deducunt*" (Boscovich 1745, § 45; I here render "*quantitas*" as "magnitude" for reasons of uniformity, even if Boscovich only later distinguished between *quantitas* and *magnitudo*). "*Ex analogia, et simplicitate naturae deducitur principium illud expositum num. 45*" (ibid., § 47).

viribus vivis, the tenet "In nature nothing happens by jump," even if variously expressed, was not identified with a single label; contrastingly, the phrase *principle* or *law of continuity* only begins to appear in 1754, in *De continuitatis lege* and in a coeval mathematical treatise added to the third volume of his *Elementa universae matheseos*.[2]

On the other hand, that Boscovich was pondering a change of approach toward continuity is suggested by a shift of emphasis in the 1748 *De materiae divisibilitate*. Here, the continuity tenet appears, but only induction is mentioned as a method capable of proving it:

> The entire force of the argument [that motions are transmitted without contact between bodies really takes place] comes from a principle accepted by many people, which is confirmed by an induction as broad as possible: in nature, nothing happens by jump, but everything that, in increasing or diminishing, comes from one magnitude to another must necessarily pass across all of the intermediate magnitudes ... *Very broad is the induction from which this principle [of continuity] is derived*, for its truth can be experienced everywhere, in any place whatsoever can it be found to be true, and no example could be offered in which it is missing.[3] (Boscovich 1757, § 66; emphasis added.)

The role of induction to justify continuity would be further specified in the 1754 *De continuitatis lege* with the distinction between complete and incomplete induction (see, in particular, Boscovich 1754b, §§ 134–135).[4] However, the remarkable

---

[2]In the following pages, I will conform to Boscovich's usage of the words and indifferently refer to the *law* or to the *principle* of continuity to signify the same content. Boscovich (1754b) seemed to prefer *principium continuitatis* when expounding others' ideas (in particular, those of the Leibnizian tradition), whereas he used *lex continuitatis* or *continui* in revealing his own point of view. This is not, however, a general pattern. In particular, Boscovich (1754b) is titled *De cotinuitatis lege* and uses expressions like "*continuitatis principium*" when quoting from or commenting upon Leibniz (ibid., §§ 3–6); however, he adopted the term "*lex*" in all other cases. The title of Boscovich (1754a) reads: *De transformatione locorum geometricorum, ubi de continuitatis lege...* Finally, Boscovich (1763, §§ 17, 31–32, and in many other places) preferred "*continuitatis lex*," and only used "*principium*" in the synopsis of the whole work (ibid., 18, referring to §§ 48 and 63). It is also worth noting that, in the 1748-drafted text of *De materiae divisibilitate*, the term "*principium*" is often associated with continuity (see, e.g., Boscovich 1757, §§ 66, 90, 91); however, the label "*continuitatis principium*" only occurs in the 1757-written preface (ibid., 133), as well as in a 1757 footnote (ibid., note *a* to § 66, 219).

[3]This is the original text: "*Argumenti vis omnis petita est a principio, et a multis accepto, et per inductionem, quantum licet, amplissimam confirmato: In Natura nihil fieri per saltum, sed quaecunque aut crescendo aut decrescendo ab una magnitudine ad aliam deveniunt, per omnias intermedias necessario transire ... Amplissimam esse inductionem illam, qua id principium colligitur, quod ubicunque ejus veritatem experiri licet, ubique verum deprehenditur, quin ullum possit exemplum proferri, in quo deficiat.*"

[4]It is not the aim of this chapter to discuss this in detail. Let us only note that a very concise definition of the two kinds of induction can be found in the above-quoted footnote to Stay's poem: "Induction [is] complete if, in proving that a property applies to all single cases, that property is generally ascribed to all of the elements of a collection; [it is] incomplete if we extend what we detect in many cases that are similar one to another—as far as no contrary instance is found amongst them—to the remaining cases of that kind, in which we have not yet been able to observe that thing. Such a principle, though cautiously applied, is certainly not infallible (I warned about, and I pointed

turn is represented in Boscovich's attitude towards the analogy of nature: in the decade between *De viribus vivis* and *De continuitatis lege*, this seems to have completely disappeared from his attempts to argue for continuity. Both in the 1748 version of *De materiae divisibilitate* and in 1754s *De continuitatis lege*, there is no further reference to the analogy of nature as being instrumental in justifying it. In the former case, the analogy of nature was mentioned in relation to many issues—but not in regard to the principle of continuity. As we have seen in the previous chapter, here, an important use of the analogy of nature was in the context of the theory of matter, both for arguing for the material points and extending their properties to the aggregates. But another employment is mentioned as well:

> It is marvellous how simple and ubiquitously unvariable from itself [*quam simplex, quam sibi ubique constans*] nature turns out to be—how clear and distinct is the overall structure of the Universe. Indeed, the extension of bodies, the impenetrability, the hardness of parts ... and the great variety of their forces, as well as looser constraints ... all that which concerns general physics, the distinction between elastic and soft bodies, fluids and solids, the universal gravity, the laws of the transmission of motions, the entire mechanics and many other, almost countless things—all of this is derived from a single principle [that is the law of forces]. No one fails to see how much stronger [this law] becomes. It is no more an arbitrary hypothesis alone, but an adequately confirmed conception, because of such a simplicity and analogy of nature [*simplicitas, et analogiae Naturae*] unvariable from itself. (Boscovich 1757, § 76)

In *De continuitatis lege*, the analogy of nature only appears in two places, but in neither case is it brought into connection with the principle of continuity. The first mention concerns the idea of the "virtual extension or divisibility" through which, according to Boscovich, Zenonists tried to prove that matter is only finitely divisible. Even if he agreed that matter was composed of a finite number of indivisible entities, he thought that Zenonism, i.e., the doctrine that real extension is composed of unextended *and* contiguous points, even in the updated form of the inflated points, was fundamentally mistaken, "since it is completely opposed to the analogy of nature and to induction from what we observe [*cum analogiae Naturae, et induction desumptae ab iis, quae videmus omnino contraria sit*]" (Boscovich 1754b, § 26). The second occurrence is more structured and sophisticated, possibly reminiscent of Newtonian transduction:

> Any absolute property—i.e., a property that does not have any relation to our senses—is generally revealed in the perceptible masses of bodies. We must transduct [*debemus*

out, the things that we should beware of in my very recent dissertation *De continuitatis lege*); nevertheless, it is extremely appropriate to research [*Inductionis ... non perfecta, ut cum a proprietate in singularibus omnibus demonstrata, eandem generaliter omnibus collectionem illorum constituentibus tribuimus, sed imperfectam, cum quae deprehendimus in plurimis similibus inter se, nullo similium contrario invento, extendimus ad reliqua ejusdem generis, in quibus nondum observare rem licuit. Principium hujusmodi caute adhibitum (quae autem cavenda sint, monui, ac demonstravi in recentissima mea dissertatione de Contiuitatis lege) est non quidem infallibile, est tamen principium investigationi aptissimum*]" (Boscovich 1755b, 46 footnote *a*). On Boscovich's notion of induction and its potential sources, see Guzzardi (2016, esp. 39–41). About induction and continuity, also see Nedelkovitch (1922, 121–126).

> *transferre*] such properties to any particle, however small ... As soon as absolute, non-relative properties are concerned [*ubi agitur de proprietatibus absolutis non respectivis*], we must consider all that we see as being common within the limits perceptible to us [*intra limites ... nobis sensibiles*], as likewise common even under such limits [i.e., in the non-perceptible realm]. And, indeed, with respect to how things are in themselves, those limits are an accident; so, if any violation of the analogy [*laesio analogiae*] should ever take place, it would much more easily fall within the limits perceptible to us, which are by far wider, than under such a threshold that is certainly near to zero" (ibid., § 135).

Summarizing, during the 10 years between *De viribus vivis* and *De continuitatis lege*, Boscovich's approach to continuity seems to have changed, so that we can distinguish two phases in the development of his reflection on this issue. In a first phase, around 1745, continuity was considered as a general tenet that did not need a profound investigation and could be proved by induction (that is, by considering examples in which the continuity of natural process is observed), as well as by a reference to the analogy of nature. In the second phase, the tenet obtained the certified label of *law* (or *principle*) *of continuity*. Such a change entailed a better and more structured justification of it, i.e., it required a new strategy of proof. This still involved induction, but no longer included the analogy of nature.[5]

In place of this, starting with *De continuitatis lege*, we now find a proof "*ex metaphysicis principiis*" (Boscovich 1754b, § 131). Years later, Boscovich would refer to that proof in terms of a novel insight in developing his perspective—something that was relatively new and not contained in the earlier presentations of his theory: beyond induction, he claimed to have "discovered [*adinveni*] a different, metaphysical argument, advanced in the dissertation *De continuitatis lege*, having derived it from the very nature of continuity" (Boscovich 1763, § 48). In the next section, I will expound upon such a metaphysical argument.[6]

[5]As far as I can see, my account on this point differs very much from the conceptual analysis by Martinović (1987). According to him, the analogy of nature was and remained the first step in Boscovich's "deductive chain" for proving his theory of forces. However, I contend that textual evidence is lacking and suggests, on the contrary, that Boscovich greatly limited the role of the analogy of nature after *De viribus vivis*.

[6]I will not examine, however, the list of phenomena that Boscovich considered in his proof by induction: As interesting as they might individually be, on the whole, they are only supposed to add (empirical) evidence. The cases are exposed in Boscovich (1754b, §§ 137–157, 1763, §§ 39–47). Note that they also include examples from geometry and the theory of numbers. The reason for this inclusion is that Boscovich considered induction as merely consisting of instantiations from all possible domains. Very interesting, in this vein, is the case made in Boscovich (1754b, § 149), where he claimed that we are used to considering numbers as discrete only because we deliberately or implicitly "disregard the intermediate quantities, that is to say, the irrational and rational numbers that fill any void between any two near numbers."

5.2 Leibniz in Light (and Shadow) of Aristotle

Whereas the continuity tenet that was endorsed in 1745s *De viribus vivis* was somewhat vague, Boscovich (1754b, § 3) seemed to suggest that the "law of continuity", even if it can have several formulations, had found a noticeable and very precise expression in the Leibnizian tradition. It is, indeed, "a very famous principle... that was articulated by Leibniz already in 1687, although he used another name for it." In Boscovich's essential reconstruction, Leibniz employed it against Descartes' laws of impact; thereafter, the principle was explained by "many Leibnizians, whose arguments would be collected by the very learned Madame De Chatellet [sic] in her *Institutions de physique*" (ibid., § 3).[7]

However, even if the appeal to Leibniz's formulation represents a new, original turn in Boscovich's development of the concept of continuum, it also appears to be strongly mediated by his acceptance of Aristotle's doctrines. The Leibnizian statement of the *lex continui* and its explication, indeed, are postponed until § 100: according to Boscovich (ibid., § 6), they required a clarification of "the nature of the continuous quantity . . . by means of geometry," and this begins with a discussion of the treatment of continuity by Aristotle—"whom Leibniz praised so much in the above-quoted essay," as he rhetorically stated.[8] A passage from the *Categories* introduced the issue:

> Discrete are number [and language]; continuous are lines, surfaces, bodies, and also, besides these, time and place. For the parts of a number have no common boundary at which they join together . . . A line, on the other hand, is a continuous quantity. For it is possible to find a common boundary at which its parts join together, a point. And for a surface, a line; for the parts of a plane join together at some common boundary. Similarly, in the case of a body, one could find a common boundary—a line or a surface—at which the parts of the body join together. Time also and place are of this kind. For present time joins on to both past time and future time. Place, again, is one of the continuous quantities. For the parts of a body occupy some place, and they join together at a common boundary. So the parts of the place occupied by the various parts of the body, themselves join together at the same boundary at which the parts of the body do. (*Cat.*, 4b36-5a13: see Aristotle 1984, I, 8; quoted in Boscovich 1754b, § 6)

Following Aristotle, if we divide a continuous quantity, we obtain two contiguous parts that are joined by a "common boundary", i.e., an indivisible point, such as A. Point A is indivisible, meaning that it cannot be counted one time as one *part* of

---

[7]Boscovich refers to Leibniz's *Lettre sur un principe general utile à l'explication des loix de la nature*... (Leibniz 1687) and to Émilie du Châtelet's *Institutions de physique* (Châtelet 1742).

[8]There is, of course, more than mere rhetoric here, since Leibniz variously appealed to Aristotle's treatment of continuity throughout his works. See, e.g., Breger (1990), Beeley (1996), and Arthur (2014, esp. 150); see also Arthur's introduction and discussion of the potential Aristotelian sources in Leibniz (2001, esp. xxix–xxxiii, 347–352). Moreover, as Arthur (2018, 276) remarks, "Leibniz's employment of [some] scholastic terms reflects his deeply held belief that many Aristotelian principles, when properly interpreted, provide a sound foundation for the mechanical philosophy that is otherwise lacking." On Leibniz's use of Aristotle's doctrines to this aim, see also Garber (2009, 5–9).

the first segment and another time as one *part* of the second. We cannot duplicate it, and we cannot divide it. Rather, it is a junction at which we pass from one segment to another.

As Aristotle proposed, we should view the language as an instantiation of the discrete. One can separate a word into its constituent syllables. In the word DISCRETE, for instance, the first syllable begins with D and ends with S, and the second begins with C and ends with E. In this case, Aristotle would have commented that "there is no common boundary at which the syllables join together, but each is separate in itself" (*Cat.* 4b 35; see Aristotle 1984, 8). This is not the case with continuous quantities. Take a line, for instance: one can find a point on it, and this is a common boundary between two contiguous segments. The point is indivisibly part of both. If an individual mentally separates the segments and uses the point two times, associating the point one time with the end of the first segment and the second time with the beginning of the second segment, she has indeed *really* treated the segments as separated, thus destroying the continuity, because the boundary is no longer common. (In fact, she has feigned dividing an indivisible point.)

This is also why continuous quantities cannot be conceived as a set of points, as we know from the previous chapter. After all, points are boundaries and must be connected by a continuous quantity. Based on this, Boscovich (1754b, § 10) emphasized the concept of boundary in Aristotle's discourse: "From the nature of the boundary also follows that a boundary cannot touch another boundary, for a continuum of which they are boundaries must always lie between them." Therefore, continuous quantities can also be conceived as those in which, in order to pass from one boundary to another, one needs to pass through any intermediate space lying between them (and, of course, when given two boundaries, there will always be an infinity of other spaces with their possible boundaries between them). Now, let us quantify the quantity, so to speak, and think of any boundary—any point—taken in a given quantity as a magnitude of that quantity. We can reformulate the previous statement as follows: *Continuous quantities are those in which, in order to pass from one magnitude to another, one needs to pass through any intermediate magnitude.*

Is there any compelling reason to maintain that natural processes are conceivable in—and only in—continuous terms? Why are we not able to conceive of, say, velocity increments or decrements as being discrete? Boscovich has a simple, smart answer to these questions: *Because any change occurring in nature happens as time passes*. Let us remember that, according to Aristotle, time offers a perfect example of a continuous quantity, "for present time joins on to both past time and future time." Any instant in the timeline is like a point on a line: it is a boundary between the past-timeline and the future-timeline. So, contrary to segments and intervals, instants are indivisible and not contiguous: the instant-boundary connecting past and future belongs both to the past *and* future timelines—it is the last "point" of the past series and the first "point" of the future series.[9]

Now, let us consider the passing of time in relation to the motion of a certain body: since any of its instants is an indivisible boundary connecting its past and its future timelines, none of them can be counted twice. However, according to Boscovich, this is the kind of mistake that we fall into if we maintain that velocity changes occur instantaneously. First, this would imply that we connect two different velocities with the same instant of time:

> If, indeed, in some instant of time, a jump occurs, in that same instant, the quantity should assume two distinct magnitudes, that is, the last one of the continuous series pertaining to the preceding time and the first one of the continuous series pertaining to the following time. For, in the same manner, as that same instant is the last one of the preceding time, as well as the first one of the following time, so the magnitude that is found at that instant should be the last term of the series corresponding to the preceding time, as well as the first term of the series corresponding to the following time.[10] (Boscovich 1754b, § 132)

Let us assume that time is a continuous quantity per se and see what happens if we consider a certain natural process as a function of time while insisting on thinking of velocity as a quantity that changes instantaneously. Let us represent time on a graph on the x-axis and the quantity changing in time (for example, *velocity*) on the y-axis. Of course, we are allowed to associate multiple values of x with a single value of y for representing uniform velocities during certain time intervals (say, velocity degree 1 with all real values of $x < 3$ and velocity degree 2 with all real values of $x > 3$). Now, if we wish to represent a change of velocity from one (continuous) time interval to another—i.e., from velocity degree 1 to velocity degree 2—through the

[9]Note that Boscovich (1754b, § 33) explicitly explains time as a continuous entity in Aristotelian terms: "Amongst the things that are continuous, we should also consider time, as we have seen from Aristotle. Time, indeed, continuously flows, and its parts succeed one another without any intermediate gap [*sine ullo intermedio hiatu*]. Thus, even in time, as in a line, we ought to distinguish between a continuous time—such as the hour—and a boundary or a limit point [*termino, vel limite*] dividing two continuous times; we shall call it 'an instant' [*momentum*]. Continuous time will correspond to line, an instant to a point. An instant will be indivisible like a point; continuous time will be infinitely divisible like a line. In the same manner as in a line, so in continuous time, there will be no particle [*particula*], however small, such as there cannot be another even smaller; nor will there be an interval of time, however large, such as there cannot be another even larger. In a determined lapse of continuous time, there will be no particle that, taken as a whole, will be the first or the last one. In any finite temporal interval, there will always be a first and a last instant, but there will never be a second and a penultimate instant. Between any two instants, however near . . . there will be a continuous time, and other instants belonging to it will be nearer to one or the other extreme, as we stated about the points of a line. As a line is generated by a flux of points (not by replicating or multiplying them), in the same manner, time is generated by the continuous duration of a thing that exists in the individual instance and lasts for a continuous time."

[10]This is the original text: "*Si enim aliquo momento temporis haberetur saltus, eodem illa quantitas binas magnitudines habere deberet, nimirum postremam seriei continuae pertinentis ad tempus praecedens, et primam seriei pertinentis ad tempus consequens. Utenim illud idem momentum et est postremum temporis praecedentis, et primum sequentis, ita magnitudo, quae habetur illo momento debet esse et postremus terminus seriei respondentis tempori praeceenti, et primus seriei consequenti tempori respondentis.*"

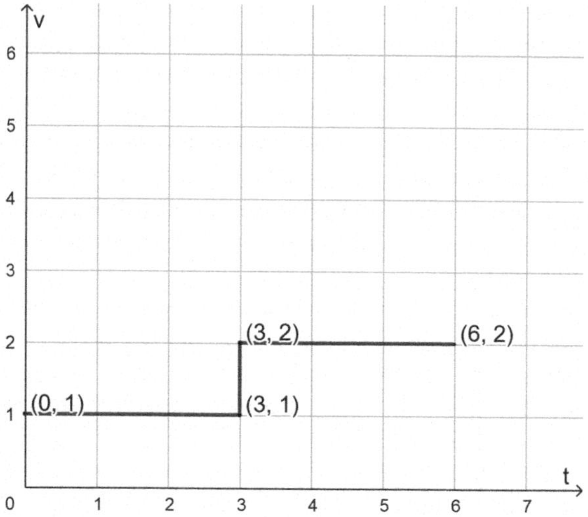

Fig. 5.1 Graph of an impossible change of velocity according to Boscovich. Taken for granted that the velocity v is a function of time t, and that time is a continuous quantity, the passage from velocity degree (ordinate) 1 to velocity degree (ordinate) 2 ought not to be discrete. Otherwise, the instant of time 3 would become associated with both velocity degrees (i.e., with both ordinates) at once. This graph is inspired by Figs. 19, 20, and 21 in Boscovich (1754b, Table I; reported in Boscovich 1763, § 51, as Fig. 5, 6, 7)

instant $x = 3$, we cannot conceive of that passage as being discrete, otherwise the same value of x (time) would become associated with multiple values of y (velocity), as if a body were to move at the same instant of time with two different velocities.[11] The diagram of Fig. 5.1 instantiates this concept.

The association of two different velocities with the same instant of time leads to another ill consequence that Boscovich calls a "duplication" or "replication" in space—that is, the same point of real space (the position of a body) becomes associated with two different points of time (instants), as if something could exist here and there at the same moment (see Boscovich 1754b, § 35, 131, 1755a, 22, 1763, §§ 49–50). Let us consider Boscovich's definition of velocity as the determination to cover a certain, determined space in a given time (the definition that he already gave in *De viribus vivis* and that now resurfaces in *De continuitatis lege*: see ibid., §§ 36, 131). If this holds true, attaching two different velocities to a

[11]In more formal terms, there does not exist a *continuous*, definite function for every $x \geq 0$ that coincides with the function $y(x) = 1$ for $0 \leq x < 3$, and $y(x) = 2$ for $x > 3$. Of course, a number of questions are left in the background and cannot be analyzed here—among other things, the tension between the notion of instantaneous velocity and that of average velocity (i.e., the relation between the space traversed and the time employed, $v = \Delta s/\Delta t$), as well as Newton's method of first and ultimate ratios and the related question of the use of infinitesimal methods in mechanics. All of these issues are implicitly or explicitly present in *De continuitatis lege* (e.g., Boscovich 1754b, §§ 36, 42–52). I am indebted to Claudio Bartocci for the discussion that we had about this issue.

single instant of time means ascribing to a body a determination to cover, in the same time, a certain determined space *as well as* another determined space—i.e., thinking that it will be found in two different positions at the same time. So, either we consider velocity as a continuous quantity—such as, in order to pass from one magnitude to another, it needs to pass through any intermediate magnitude—or we unpleasantly fall into aporias. (Finally, note that, in as much as the whole argument avoids any reference to empirical conditions, only coming from the discussion of the Aristotelian notion of continuity and its application to continuous quantities, it is a metaphysical argument.)

This emphasis on connected couples of variables, such as *time* and *velocity* in the example given above, was not included in Boscovich's early statements on the principle of continuity, but it is the distinguishing feature of his view as expressed in *De continuitatis lege*. It is not clear whether he arrived at his more articulated conception by being inspired by Leibniz or if his background Aristotelianism drove him to embrace (and partly reformulate) Leibniz's *lex continui*. Maybe the one possibility does not exclude the other. In any event, all of this reveals Boscovich's attitude toward Leibniz:

> [He] proposed [the continuity law], without calling it by that name, in the above-mentioned article, which states the following . . .: "When the difference between two instances in a given series or that which is presupposed can be diminished until it becomes smaller than any given quantity whatsoever, the corresponding difference in what is sought or in their results must of necessity also be diminished or become less than any given quantity whatsoever." And also: "When two instances or data approach each other continuously, so that one at last passes over into the other, it is necessary for their consequences or results (or the unknown) to do so as well." Finally, more generally: "As the data are ordered, so the unknowns are also ordered." (ibid., § 100)[12]

Let us remember that Leibniz originally called this statement a "principle of general order". According to Boscovich, this is somewhat redundant if it is taken as the statement of the principle of continuity. As he remarked, "there is no better term than the passage [*transitus*] through all intermediate quantities in order to express" what Leibniz referred to in terms of "ordered disposition, continuity, and diminished

[12]This is Boscovich's Latin translation of the original French text of Leibniz (1687): "'*Cum differentia duorum casuum potest diminui infra quamcumque quantitatem datam in datis vel in eo quod positum est, oportet ipsa possit inveniri imminuta infra quamcumque magnitudinem datam in quaesitis vel in o quod resultat.' Tum idem sic: 'Cum casus (vel id quod datur) accedunt ad se invicem continuo ac desinit tandem unus in alium, oportet consectaria vel eventa (vel id quod postulator) idem praestent.' Demum idem generalius: 'Datis ordinatis etiam quaesita sunt ordinata.*'" Leibniz's original text reads as follows: "'*Lorsque la difference de deux cas peut estre diminuée au dessous de toute grandeur donnée* in datis *ou dans ce qui est posé, il faut qu'elle se puisse trouver aussi diminuée au dessous de toute grandeur donnée* in quaesitis *ou dans ce qui en resulte', ou pour parler plus familierement: 'Lorsque les cas (ou ce qui est donné) s'approchent continuellement et se perdent enfin l'un dans l'autre, il faut que les suites ou evenemens (ou ce qui est demandé) le fassent aussi.' Ce qui depend encor d'un principe plus general, sçavoir:* 'Datis ordinatis etiam quaesita sunt ordinata.'" In my translation of Boscovich's Latin rendering, I used Loemker's edition of Leibniz's *Philosophical Papers and Letters*: see Leibniz (1687, 744/52/351).

difference beyond any given boundaries."[13] However, Leibniz's statement may help explain how continuity works. So, Boscovich tried to re-formulate it as follows:

> Two variable quantities, or quantities that can change their magnitude, are connected one to the other so that, if the magnitude of one of them is determined, the magnitude of the other one is determined as well. Let us conceive of two magnitudes of the first quantity and two corresponding magnitudes of the second quantity; if the first quantity goes [*abeat*] from the first magnitude to the second magnitude by continuous change, passing through [*transeundo per*] all intermediate magnitudes, the same applies to the second quantity as well. (ibid., § 102)

Boscovich may or may not be mistaken in this re-statement of Leibniz's principle; however, according to it, if one of two connected variable quantities continuously passes from one magnitude to another, this "order" becomes superimposed on the quantity resulting from their connection. (As a corollary of this, such a resulting quantity, which Leibniz would call the *quaesita* or "*ce qui est demandé*", cannot assume more than one magnitude at once; that is, every value of the first quantity ought to be associated with one single value of the second quantity.) Let us take into consideration two such connected quantities, and let time be one of them: the *data* ("*ce qui est posé*" or "*ce qui est donné*" in Leibnizian terms). Now, as we have seen, Boscovich considers time as a continuous quantity *per se*. So, if we connect another quantity to it, such as velocity, and study how it changes, we are forced to recognize, by opportunely substituting in the above-quoted passage, that *if* time *goes from the first* instant *to the second* instant *by continuous change, passing through all intermediate* instances, *the same applies to* velocity *as well.*[14]

---

[13]"*Transitus autem per omnes intermedias magnitudines omnium optime exprimit* ordinationem *illam*, continuitatem, imminutionem differentiae *infra limites quoscumque datos*" (Boscovich 1754b, § 102; emphasis in the original text). Note that the expressions emphasized respectively refer to the following expressions in the text of Leibniz translated in § 101: "*Datis ordinatis etiam quaesita sunt ordinata*; *cum casus accedunt ad se invicem continuo*; *differentia duorum casuum potest diminui infra quamcumque quantitatem datam.*"

[14]Of course, the discourse on continuity does not exhaust Boscovich's incursions into Leibnizian philosophy, even if this does not necessarily imply a kind of Leibnizianism. Although many scholars have linked Leibniz's conception to Boscovich's in any number of ways, as discussed in Sect. 3.4.1, source-based studies in this realm are still lacking. Such an approach goes well beyond the aims of the present book, but the evidence that I discussed in the previous chapters seems to indicate that Boscovich's interest in Leibniz's doctrines (or in doctrines associated with his *milieu*) mainly involves three domains relevant to his natural philosophy: the notion of living forces (examined most of all in Boscovich 1745, then in Boscovich 1755b, 373–381), the monads as the constituents of matter (discussed in Boscovich 1757, § 12 footnote *a*, then in Boscovich 1763, § 139), and, of course, the principle of continuity. Note that he only endorsed the latter concept, while rejecting both the notion of living force and that of monadology (although this point certainly needs further research). However, other issues should probably be added to the list of Leibnizian concepts that Boscovich examined (and rejected). Mostly discussed or mentioned in the Supplements to Stay's Newtonian poem and occasionally reprised in later works, they at least include pre-established harmony (Boscovich 1755b, 13 note *a*, and 279) and the principle of sufficient reason (Boscovich 1755b, 14–15 note *b*, and 279–283). On this, see Baldini (1992b, 42).

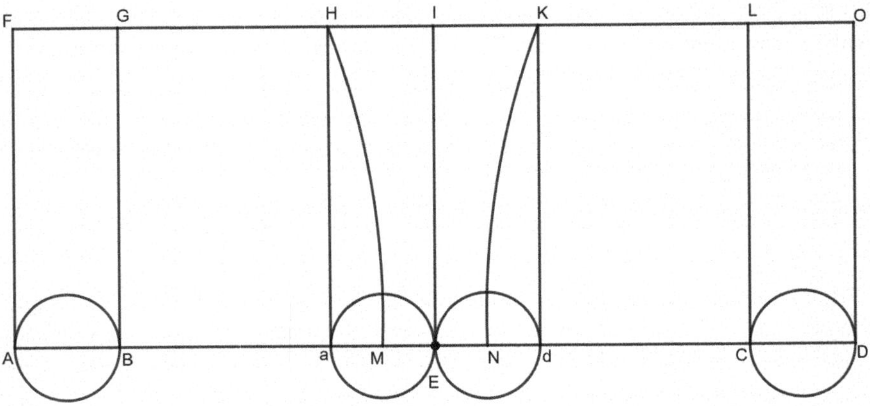

Fig. 5.2 Diagrammatical representation of the supposedly instantaneous annihilation of velocity according to Boscovich (Modern adaption from Table I, Fig. 9, of Boscovich 1745)

5.3 Problem-Solving by Geometrical Means

Perhaps his regard for the reference to the above-quoted passage from Aristotle's *Categories* as a crucial point led Boscovich to emphasize of the role of geometrical exemplification in revealing his ideas about continuity. Or, rather, it was his mathematical background that guided him towards seeing an appropriate framework for continuity in that Aristotelian passage. After all, whereas, according to Aristotle, there is no difference between mathematics and physics in terms of the objects that they investigate (only the modes of investigation are different), when commenting upon the *Categories*, Boscovich rendered the Greek term *σῶμα* ("*corpus*" in Du Val's Latin translation that he quotes) as *solidum*, which only refers to geometry (see Boscovich 1754b, § 8).

As we have seen in Sect. 2.3, in *De viribus vivis* (1745), he had employed traditional graphs in order to calculate quantities such as *momentum* compared with *living force*. Their meaning, however, went well beyond the mathematical magnitudes that they were supposed to compute, for they show that both measures of force (mv and mv^2) are legitimate. However, Boscovich (1745, § 47) also used graphs in order to schematize real situations, as in the case of colliding bodies referred to in the Introduction. Let us reproduce and inspect that figure from a more formal point of view (Fig. 5.2).

Boscovich's schematization can easily be understood: there are two equal spheres, which we represent based on their diameters, $AB = CD$. They collide at point E by going along the same line AD with equal velocities but in opposite directions. In purely mathematical terms, the model is somewhat rough and ambiguous. For example, line AD is associated with spaces, but it cannot possibly be regarded as the x-axis of a diagram. Moreover, time plays a role in Boscovich's description of the graph, but it is not represented in it. However, the diagram is not designed to show *how* velocity changes in bodies' collision, but instead aims to initiate discussion regarding *whether* the causal mechanism governing motion changes in the collision of bodies is physical impact. In other words, Boscovich's graph schematizes the following question: what if we assume that, when bodies collide, impact actually takes place—i.e., that bodies actually make contact?

As a representative of the Jesuit physico-mathematical tradition, Boscovich tended to exploit the veridical power of mathematics (of geometrical constructions, in this case). Let us assume that mathematics always grants that the truth of the premises is preserved in the chain of derivations. Hence, if we derive an implausible consequence from a certain premise, the premise itself was ill to begin with. Now, as Boscovich explained, the assumption of real impacts during bodies' collision entails that we think of velocity as a quantity that can change by jumping from one magnitude to another. After all, based on the diagram, "in *I*, every expression by means of the ordinates breaks down, and the ordinate *EI* will be followed by a point," i.e., by a null velocity. On the other hand, if we want velocity to be—as it is—a continuous quantity, we should maintain that "whenever the particles *B* and *C* were to arrive at a very small distance, some repulsive force would push them back endlessly, so that their velocities would gradually be extinguished before contact." In sum, Boscovich was not actually representing physical quantities here; instead, he was discussing a physical question ("Does physical impact really take place?") based on a metaphysical assumption ("Natural processes are continuous processes") represented through a geometrical diagram.

This representational attitude by means of geometrical diagrams becomes radicalized in the 1754 treatise *De continuitatis lege*, as soon as the law of continuity finds explicit formulation in his discourse. Here, after commenting the *Categories*, he goes on exploring continuity through an impressive number of examples from geometry, which are provided along with diagrams in the final tables of the book (see esp. Boscovich 1754b, §§ 7–15, 53–99). Of the 174 paragraphs that form the treatise, more than one hundred have geometrical content, often referring directly or indirectly to the diagrammatic apparatus in the tables at the end of the treatise.[15]

On the other hand, Boscovich could hardly have considered the very *notion* of continuity as geometrical in itself—as much as it cannot be physical in itself. As a metaphysical notion, it must lie behind both geometry and physics and affect the former as well as the latter (and, after all, geometry and physics are subalternate disciplines to the subalternating metaphysics). Therefore, he was quite serious when he stated that "geometrical continuity is not necessary in order to defend that of the physical; I used it as an example in confirmation of a wider induction" (Boscovich 1763, § 43). In other words, he claimed to have employed geometry as a collection of instances in which continuity finds an obvious application. In fact, he seemed to think that, in geometry, continuity is ubiquitous and dominates everywhere (particularly because geometry deals with the imaginary, continuous space), whereas in physics, we are not only faced with continuous processes, but are also faced with the

[15]See Boscovich (1754b, §§ 11–15, 22–24, 28–32, 53–79, 80–99, 107–123, 139–150, 155).

discrete structure of matter, even if this is a consequence of force being a continuous function of distance.

Thus, geometrical continuity does not provide grounds; it instead offers *examples*, *illustrations*, and *visualizations*, as far as we are dealing with natural processes. When used in this manner, Boscovich argued, geometry can even help in a "metaphysical discussion" of the law of continuity (Boscovich 1754b, § 131–132, 1763, §§ 32–35, 43–44). To limit the analysis to Boscovich (1754b), most geometrical instances are introduced or addressed through expressions that refer to vision and visualization. Phrases such as *ope geometria illustrare*, *oculis obicere/subicere*, and *oculis intueri* are employed. What follows is a complete list of occurrences of the word "*geometria*" in Boscovich (1754b) in which he mentioned why he decided to illustrate his ideas through diagrams:

> § 6: "In order that the notion of the principle of continuity results as being more perspicuous, in the first place, we shall pursue the nature of continuous quantities and illustrate them by means of geometry."
>
> § 19: "The infinite divisibility of extended continuum ... in geometry is actually proved by countless arguments, so that there is no room left for any doubt. It is, so to speak, brought under the eyes by the relation of incommensurable quantities, which no [finite] number can represent, by the angles of contingence intersected by bigger and bigger arcs of a circle, by the asymptotic legs of curves, indefinitely extended."
>
> § 39: "It will be allowed for us to observe all of this in a certain way with our eyes themselves by means of geometry."
>
> § 41: "By the help of the same figure [reported as Fig. 3 in the table closing the work], even the main difficulty that caused the ancient philosophers to oppose continuous motion easily disappears. This emerges from the case of the motion of Achilles and the tortoise ... First, we shall present, in whatever manner. that famous difficulty, then we shall solve it solely through the nature of continuous quantities and, finally, by invoking the support of geometry, we place the solution itself under the eyes." Once he has discussed the Achilles paradox at length, in § 46, Boscovich comments upon the solution, exploiting his Fig. 3: "If one wishes to place under the eyes that same solution, it suffices, as in Fig. 3 ..."
>
> In § 53 & ff., Boscovich investigated the concept of continuity by explicitly using curves (geometric loci): "In order to inspect better the nature of continuity itself and all that we will say about the law of continuity, it will be extremely advantageous to consider very carefully the course of a type of continuous quantity, that is, the line. This kind of consideration involves the entire higher geometry, but we shall summarize and only discuss the most useful cases. First, in geometry, there is an infinite number of continuous lines that are also called geometric loci ..."
>
> § 85: "In order that the thing may become more evident by placing it under the eyes with the help of geometry, let us, in Fig. 13 ..."[16]

[16]Here is the original text of the quoted passages: § 6: "*Ut autem ipsius principiis Continuitatis notio evadet magis perspicua, continuae quantitates naturam persequemur primo loco, & eam Geometriae ope illustrabimus*". § 19: "*Continui extensi divisibilitas in infinitum ... in Geometria innumeris sane argumentis ita evincitur, ut nullis dubitationi locus supersit. Illam incommensurabilium quantitatum relatio nullis subiecta numeri, illam anguli contactus a majorum circolorum arcubus perpetuo secti, illa asymptotica curvarum crura indefinite producta oculis fere ipsis objiciunt*". § 39: "*Licebit haec omnia ipsis quodammodo intueri oculis, ope Geometriae ...*" § 41: "*Ac ope figurae [Fig. 3] ejusdem facile etiam dissolvitur praecipua difficultas, quam contra motuum continuum Veteres objiciebant, petita a motu Achillis, & testudinis .... Exponemus primo*

In the relatively new context of Boscovich (1754b), the visualizing power of the diagrams is granted by the fact that the law of continuity, grounded in the Aristotelian notion of continuous quantities, holds well in both nature and geometry. Two different but integrating aspects are at play here—one deals with the contents and the other concerns the formal. In regard to the contents, Boscovich not only established the notion of continuity itself, but also demonstrated interest in how continuous quantities may be generated through the connection between one quantity and another quantity that is known to be continuous. For, according to his interpretation of Leibniz's continuity law, given a certain relationship between one quantity (let us call it the *quantitas data*) and another one (the *quantitas quaesita*), if a *quantitas data* is continuously changed so that it passes from an initial magnitude to a final magnitude passing through all of the intermediate degrees, it turns out that the *quantitas quaesita* is continuously changed as well. For example, if we study force as a function of distance, provided that space is a continuous quantity and that one and only one force intensity corresponds to any single distance, the *quaesita* ("Variation of force according to the distance") will also be continuous.

On the other hand, as regards the formal aspects, the diagrammatic strategy allows one to geometrically operate on the contents (the connections), so that the consequences of the diagrams are consequences of the contents themselves. It is usually convenient, as Boscovich (1754b, § 107) suggested, to think of a connection between "two quantities only, one of which is simple or can be considered as such" (where *simple* means *without further connections*), as with our example of force as a function of distance. In this case, as soon as the quantities involved are continuous, the diagrammatic representational strategy allows for their connection to "always be revealed through lines by arbitrarily assuming an axis and representing in it the one quantity through the abscissas computed from a given point, the other through ordinates inclined by a given angle. The vertex between them, as far as they continuously prolong themselves along the axis, describes a continuous line referring to such a connection."[17]

In other words, provided that the law of continuity is metaphysically grounded, we can *use* geometrical tools (more precisely, curves) in order to *represent* processes

---

quidem utut notissimam difficultatem, tum eam e sola continuae quantitatis natura dissolvemus, ac demum Geometria in subsidium vocata, oculis ipsis eamdem subjiecimus solutionem." After discussing the paradox, he commented on the solution using Fig. 3 (§ 46): "*Si solutionem eandem libeat oculis ipsis subjicere, satis est in Fig. 3 ...*" In § 53 & ff., Boscovich discusses continuity by using curves: "*Interea ad ipsam continuitatis naturam melius perspiciendam, & ad ea, quae de lege continuitatis dicturi sumus, proderit plurimum considerare aliquanto diligentius ductum unius generis quantitates continuae, nimirum lineae, quae consideratione universam secum sublimiorem Geometriam trahit .... In primis in Geometria sunt infinita linearum continuarum genera, quae etiam locos geometricos appellant ...*". § 85: "*Ut res evidentior evadat ipsis oculis ope Geometria, subjecta, sit in fig 13 ...*"

[17]In his own words: "*Earum nexus in eo casu exponi semper potest per lineas assumendo axem ad arbitrium et in eo rapresentando alteram quantitatem per abscissas a dato puncto computatas, alteram per ordinatasi psi in dato angulo inclinatas, quarum vertex in earum continuo excursu per illum axem describat lineam continuam eiusmodi nexum referentem.*"

or concepts in which continuity is involved. The discrete structure of matter is no exception, for it is suggested by the continuous curve of forces and represented by its behavior at the asymptotes (I will come back to this in commenting upon the mathematical features of the curve in Sect. 6.5). But mathematics, so to speak, is language, not substance. In using it as an adept of the *physico-mathesis*, Boscovich only appealed to its representational virtue, without being committed to any ontological option regarding the relation of mathematics to physics or other disciplines.

Chapter 6
Touching Infinity

6.1 Early Expressions

It was probably the new view of continuity, together with his enhanced diagrammatic strategy, that led Boscovich to ultimately re-frame the law of forces and search for its firm mathematical basis. However, it was a process that took approximately a decade of profound exploration of a kind of mathematics with which he was, at least at the beginning, less familiar.

Boscovich drew his force curve for the first time in the 1745 dissertation on living forces. In order to avoid abrupt velocity changes in collision, as well as in other physical phenomena, at every distance, some repulsive or attractive "action" must take place, and this should be expressible "through ordinates to certain continuous curves" (1745, § 48), so that a graph can be formed as follows:

> [In Fig. 6.1a], let segments of the straight line *AG* display the reciprocal distances of two particles, and let there be a certain curve *MCKIH* whose nature is such that it has the straight line *NL* as an asymptote perpendicular to the axis *AG*, from which it perpetually recedes. Let this curve intersect the axis at some point *C* from which it recedes until reaching *K*; then, let its direction turn backwards and, from some point *I*, have, as an asymptote, a second-degree hyperbola such as it is under the abscissas and is constant with respect to the square of the ordinates ... Let the ordinates *BM* of the curve oriented towards the other region [*plaga*] express the repulsive force; let those oriented towards the opposite region—like *DK*, *EI*, *FH*—express attractive forces. (Boscovich 1745, § 50)

Figure 6.1a embodies the general idea for avoiding jumps by means of a force-inversion. However, Boscovich argued that, if we conjecture that all phenomena are conceivable in those terms, we have to take into account the possibility of a greater number of force-inversions as we determine the scale of distances involved and allow our description to become more precise. "Curves with a much greater number of going and coming of the line [*itus curvae, & reditus*]" will then be formed (ibid., § 56), similar to that represented in Fig. 6.1b (and commented on in ibid. §§ 54–55).

L. Guzzardi, *Ruggiero Boscovich's Theory of Natural Philosophy*, Science Networks. Historical Studies 60, https://doi.org/10.1007/978-3-030-52093-9_6

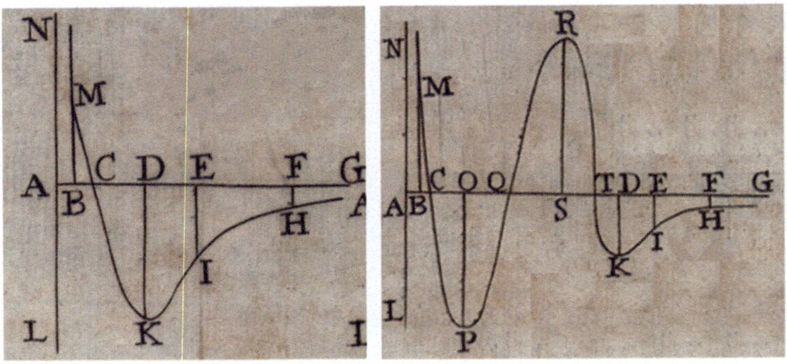

Fig. 6.1 (**a**) (left) and (**b**) (right). Boscovich's first two versions of the curve of forces (From Figs. 9 and 11, respectively, Table I, in Boscovich 1745)

Nevertheless, in 1745, the task of a complete and unitary theory of forces seemed too broad and ambitious to him, or perhaps simply premature:

> To follow in detail [this going and coming of the curve] would be an infinite task, and this is not the appropriate place to do this. But let us notice that all such curves are of the type called parabolic [*polynomial* in modern terms]; they are of the form $a + bx^m + cx^n + dx^r + ecc. = y$. That is, given a distance x, there is a single force y that is either attractive or repulsive; but the same force y may correspond to many distances x. Since Newton has already solved the problem of finding a curve of a parabolic type that passes through any given number of points, a continuous and regular curve can always be found, so that it expresses forces of whatever particle relative to whatever other particle, and such forces are deduced from phenomena. Moreover, the same curve can approximate any given arc of any other curve and intercept them in any number of points as close as they might be, provided that those arcs correspond to different portions of the axis. However, as regards the nature of these curves and the points that they pass through, they must be investigated from the phenomena. (ibid., § 56)

Boscovich referred to Newton's (1687, 896–897) Lemma V in Book III of the *Principia*: "To find a parabolic curve that will pass through any number of given points." However, such parabolic curves could hardly serve his purposes, since they do not have vertical asymptotes. As such, either Boscovich was speaking in very generic terms and only expressing that his curve of forces mathematically exists (for, given any number of points, a curve passing through them can always be found) or he was simply being inaccurate for some reason. In any case, the reference to Newton's Lemma V contradicts the requirement of the above-quoted § 50 from *De viribus vivis*, where *NL*, as drawn in Fig. 6.1a, is said to be an asymptote of the curve. Moreover, in Boscovich's theory, the asymptotic behavior is essential for two correlated reasons: the interactions at infinitesimal distances and the discrete structure of matter. (I will return to the mathematical aspects of the asymptotes in the following pages.) And, in any case, the mention of "parabolic curves" disappears in the later presentations of the law of forces.

A breakthrough regarding the mathematics underlying it occurred in *De lumine* (Boscovich 1748), wherein he accompanied the qualitative presentation of the curve with some general mathematical conditions. He did not reach the mathematical

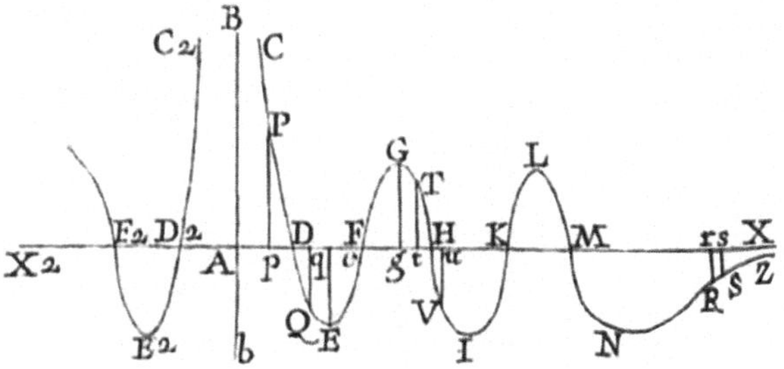

Fig. 6.2 The third version of Boscovich's curve of forces with a negative branch (From *De lumine pars II*, 1748)

confidence and precision that he later displayed when providing the analytical expression of the curve (Boscovich 1755a), but the first condition expressed in *De lumine, pars II*, already clarified that *bAB* (see Fig. 6.2) is an asymptote of the curve such that "both legs *DC*, tending to repulsive parts, must approach that asymptote. They must be perfectly equal [*et similia prorsus et aequalia*] and the area *BADC* enclosed by the axis, the asymptote, and the asymptotic leg [*crure asymtpotico*] must be infinite" (Boscovich 1748, § 6).[1] The resulting graph is Boscovich's third attempt to draw a curve of forces, and it is practically its final form. (Let us compare it with the graph of the curve in the *Theoria*: see Fig. 2 of the Introduction.)

Both features—the "perfectly equal" negative branch indefinitely prolonging into the repulsive sector and the infinite area *BADC*—have remarkable consequences, which I will examine in the following sections. In *De lumine*, Boscovich did not further analyze the mathematical underpinnings of the curve, but at the time he could hardly have ignored that these conditions are satisfied by hyperbolas. As he emphasized in *De continuitatis lege*, if the force is expressed as a function of distances, the area of the subtended curve must be proportional to the square of velocity (see Sect. 3.4). Then, in order that the repulsive force is able to destroy any velocity, however great, the area of the curve must be infinite. This requires that "the force decreases at least as the simple ratio to the distance: that is, in fact, the ratio expressed by a conic hyperbola . . . In the whole family of the hyperbolas, all those whose ordinates grow less than this have a finite area; those whose ordinates grow more have an infinite area as well, but infinitely greater than infinity" (Boscovich 1754b, § 164. I will return to this issue in Sect. 6.5.5 of the present chapter.)

Therefore, we should conclude that, at a certain point between 1745 and 1748, Boscovich realized that the analytic expression of the curve of forces would not be of the "parabolic kind" or a polynomial. Instead, as it would ultimately become clear in his 1755 work, it was a ratio between two polynomials: i.e., a curve of a hyperbolic kind. This was probably the effect of his increasing acquaintance with Calculus and

---

[1]It is unclear whether, when describing the two legs as "*et similia prorsus et aequalia*," Boscovich aimed to account for the qualitative (*similis*) and the quantitative (*aequalis*) aspects or if this qualification is merely an emphatic expression. *Crus*, which I render here as *leg*, is an intuitive notion that refers to one extreme arc of a curve near an axis. For example, the asymptotic leg of a hyperbola is an arc near the asymptote, but a parabolic leg is the arc of a parabola starting from the origin of the axes.

signaled the growing importance of mathematical aspects, for him, in formulating his natural philosophy. The 1755 dissertation *De lege virium in natura existentium* and its reprise as a supplement to the *Theoria* (see Boscovich 1755a, 1758, Suppl. I, 1763, Suppl. III), including the analytical expression of the curve, were the final stages of this development.

6.2 Infinite Legs and Their Arcana

In Sect. 4.1, I referred to the circumstance in which, in 1747–1748, Boscovich was projecting a broader work in natural philosophy, which I have called *the Camaldolese Ur-Theorie*, in which geometry should play a certain fundamental role. In an above-quoted letter (September 14, 1748), he also informed his brother Natale about this project and about the dissertations that he was drafting. He remarked that he had exposed the theory in *De lumine, pars II*, defended some days before, but "in the big book there will be a little more of substance as far as physics is concerned, yet there will be much more on Metaphysics, Geometry, and Calculus."[2] It is easily conceivable, on the other hand, that Boscovich, as *the* mathematician of the Roman College, spent much of his time absorbed in mathematical research and its various aspects, and this finally came to intersect with his ongoing project on natural philosophy.

Let us remember that, in the same letter to Natale, he recounted his recent replies to some mathematical queries. One of them, in particular, concerned curves and led him to prepare "a dissertation of some pages" that he planned to (but ultimately did not) publish. The dissertation might be preserved as a manuscript in the Boscovich Archives; in any case, his interest in the curves transpired from the coeval work *De maris aestu* (Boscovich 1747). As we have seen (Sect. 1.7), in the final sections of this treatise, Boscovich explained his idea of a "sidereal space" for allowing the ideal absolute fixity of the earth in a Newtonian context. He also discussed basic notions of mechanics, such as uniform rectilinear motion and accelerated motion, which he brought in connection with his mathematical speculations about curves. In particular, since Boscovich considers the principle of inertia to be an empirical law, he argued that—against Euler's proof of inertia *via* the principle of sufficient reason—there are no *a priori* reasons why a body left alone, with no forces acting upon it, should follow a rectilinear path.[3] However right or wrong he was, he seemed

[2]"*Nella seconda [parte di De lumine] vi è abbozzata tutta la mia teoria da principio. Vi sarà nell'opera grande poco più di sostanza in materia puramente fisica, benche vi sarà assai più, di Metafisica, di Geometria, e di Calcolo*" (Ruggiero Boscovich to Natale, September 14, 1748, in Boscovich 2012b, I, 167).

[3]Boscovich's criticism is revealed in his (1747, §§ 86–90). The complaints mainly concern (and point out the weaknesses of) Euler's *Mechanica*, which is duly quoted, together with other passages (see esp. Euler 1736, §§ 56–65, 27–29). The following citation (from Euler 1736, § 56, demonstration of Proposition 7, quoted in Boscovich 1747, § 86) is particularly emphasized in Boscovich's

to think that Euler's demonstration was based on an argument of simplicity. On the contrary, he claimed that "the straight line is one of infinite lines, and all lines—however they come back to themselves, however they bend one with respect to the other—are in themselves equally simple as the straight line" (ibid., § 90) . He contended that the consideration of the straight line as the simplest line is merely an effect of poor mathematical education. "Besides," he continued, "the straight line itself involves a notion of infinity that by far exceeds the power of the human mind." As such, it required special research upon which Boscovich was working at the time:

> This topic needs a whole and certainly more extensive dissertation, which we, among many other things, set aside for our elements of the conic sections [*Sectionum Conicarum elementis*], already arranged for the most part and only requiring the finishing touch. There, we will expound upon the astonishing nature, the astonishing transformation, and the connection [*nexum*], as well as the arcana of the infinity [*Infiniti arcana*] that are absolutely necessary, if infinity is admitted, but exceed by far any capability of human understanding. (ibid.)

In fact, the above-mentioned "elements of the conic sections" form the third volume of his coursebook of mathematics. Probably, for that time, he only employed informal notices for lecturing, since the first two volumes were published 5 years later, and the new edition, ultimately including a tome concerning *Sectionum conicarum elementa* ("the elements of the conic sections"), was not published until 1754.[4] Yet, the 1747 dissertation *De maris aestu* enables us to take a closer look at the work that was in progress in Boscovich's mathematical cabinet:

---

treatise: "A body in an absolute state of rest must persevere in such a state unless it is disturbed to move by an external cause. *Demonstration.* We conceive that this body exists in infinite and empty space, and it is evident that there is no reason why the body should move in one direction rather than in another. Consequently, because of the lack of a sufficient reason why it should move, it must remain at rest forever . . . And indeed, it cannot be believed that, in that empty infinite space, the lack of a sufficient reason to perform a motion is the only cause for a body to remain at rest; but there is no doubt that the cause of this phenomenon can be found in the nature itself of the body. Clearly, the lack of a sufficient reason cannot be taken as the true and essential cause of an event whatsoever, but it only rigorously shows the truth [i.e., what actually takes place]. Indeed, it signals, at the same time, that the true and essential cause is concealed in the nature of the thing itself, and this does not cease when the lack of a sufficient reason ceases."

[4]Boscovich's coursebook of elementary mathematics has a complicated history. It was first published in 1752 in one "tome" consisting of two different "parts" (volumes), each of them bearing the convoluted title of *Elementorum universae matheseos ad usum studiosae juventutis, Tomi primi Pars prima* and *Tomi primi Pars Altera* (that is, more or less, *The First Part of the First Tome of the Elements of General Mathematics to be Used by the Students*, and *The First Part of the Second Tome, etc.*). The first part details the geometry of the plane and solid figures, arithmetic (including the logarithms), and trigonometry, both plane and spherical; the second part treats the finite algebra. As mentioned, the edition from 1754 added a third volume, or Tome III, on the conic sections (which also included a treatise on geometric loci that I will examine in the following pages), whereas the "two parts of the first tome" composing the 1752 edition are now presented as Tome I and II under a slightly different and more concise heading (*Elementorum universae matheseos . . . Tomus I*; . . . *Tomus II*). Pepe (2010) provides a detailed reconstruction of the editorial history of this work.

> The straight line by its own nature goes to infinity on both sides and has no limits. Considering a straight line as bounded amounts to considering a segment of a geometric locus, precisely in the same way as considering an arc of a circle amounts to considering a segment of its circumference. An infinite straight line in itself is then similar, in a certain way, to the circumference of an infinite circle, which, in a certain way, comes back to itself at infinite distance. So, [a straight line] is connected to itself [*conjungitur*], as also happens with the legs of a parabola or wth the opposite branches of a hyperbola, and the same is actually true for the infinite legs of any geometric locus. All of these do not break off anywhere and, if infinity may be admitted, are connected and joined to each other at an infinite distance and, in most cases, with mutually opposite directions: thus, a parabola and even the two branches of a hyperbola are not different from a whole and continuous ellipse, provided that certain mysteries of infinity are duly understood and handled. (Boscovich 1747, § 90)

In the subsequent years, he must have realized that more time and a certain number of pages were needed for an exhaustive examination of such topics. Hence, the projected "more extensive dissertation" about the infinite quickly lengthened and was finally added to the third volume of his coursebook in 1754. The result was a long treatise consisting of 172 *in-quarto* pages under the headline, *De transformatione locorum geometricorum, ubi de continuitatis lege, ac de quibusdam Infiniti mysteriis*, i.e., "On the transformations of geometric loci, where it is dealt with the law of continuity and certain mysteries of the infinity" (Boscovich 1754a). On the one hand, the title encapsulated the research, announced in 1747, on the notion of infinity and its "mysterious" aspect, which "exceeds the power of the human mind"; on the other hand, it emphasized and connected this issue with Boscovich's more recent interest in the philosophical underpinnings of the *lex continui*.

In the 1754 treatise, he sought to detail all possible kinds of transformation, from one geometric locus to another, in accordance with 11 "canons" or general rules (Canon I is stated in § 764, and Canon XI is stated in § 862).[5] In this new context, the statement mentioned in his 1747 work (§ 90), i.e., that the infinite straight line is like the circumference of a circle having infinite radius, is raised to the rank of one such rule, the Canon X (Boscovich 1754a, § 858). Infinity can now be viewed as the property, possessed by all geometric loci, to "come back" to themselves. And, such a feature being embodied in a canon, it can be used in a variety of "transformations".

A paradigmatic case is constructed and displayed by means of a diagram in Boscovich (1754a, § 716 and the related Fig. 264 of Table VII), which is reproduced in simplified form in Fig. 6.3. (Unfortunately, Boscovich's figures are often somewhat complicated, since they are usually employed for many diverse purposes; i.e., each one can serve as a "basis" for a number of constructions to be used in several demonstrations.) Given point C on straight line *r*, let a circle with center C be drawn and let the straight line rotate on it in a given direction, such as counterclockwise. We intend to study its intersections with the circle and with another straight line, *s*, at a certain definite distance, beginning with intersection P_0 in a perpendicular position. Any point of intersection P_0, P_1, P_2, P_3, $P_{4\ldots}$ between *r* and *s* has a corresponding

[5] On Boscovich's methods of transformations, see Manara and Spoglianti (1979), Martinović (1991, 2015), and Del Centina and Fiocca (2018a, b).

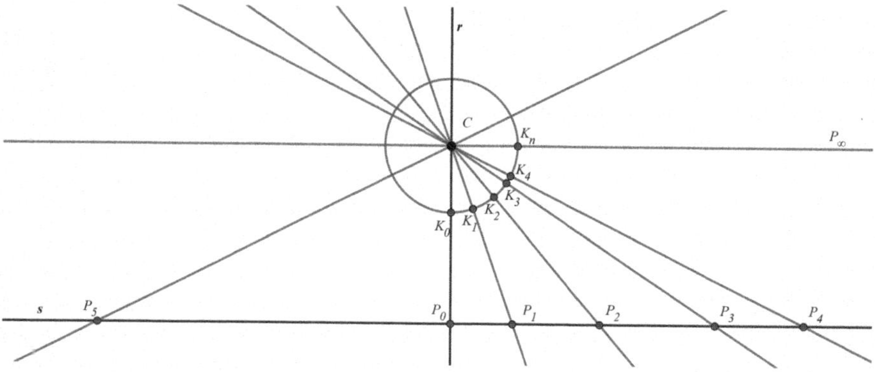

Fig. 6.3 Boscovich's idea of the "transit through infinity" and the consequent "come-back" (simplified graph, adapted from Boscovich 1754a, Table VII, Fig. 264)

point K_0, K_1, K_2, etc., on the circumference, and vice versa. By continuing to rotate *r*, the line will get to intersection K_n with the circle, and *r* and *s* will be parallel. Where will the corresponding intersection (e.g., P_∞) be on *s*? In this case, Boscovich argued, the intersection between the straight lines "will be nowhere, as it was absorbed into that immense sea of the infinite." However, a paradox arises from Boscovich's perspective: on the one hand, the two lines are parallel, so *r* "will never meet [*s*], even if it were to be continued to infinity"; on the other hand, let us attempt to abstract from the intersection K_n and just consider the subsequent intersections, until *s* gets to the point P_5 on the other side of *s*. We are led to the conclusion that intersection P_∞, "after departing for infinity [*post discessum in infinitum*], will conceal itself [*delituerat*] in the only instant of time in which [the intersection with the circle was in K_n] will be immediately found on the opposite side," such as in P_5. Hence, as Boscovich (1754a, § 717) commented, "Such transit of the point P through the infinity from a region to the opposite one seems to be accomplished by a motion that is absolutely continuous, as if, at that infinite distance, the one infinite [half-line] were, in a certain manner, connected to the other infinite [half-line]."

This is an instantiation of a general approach, which finds ubiquitous application to conic sections and their mutual transformations in the 1754 treatise *De transformatione locorum geometricorum*.[6] An overview of various kinds of

[6] Boscovich's perspective implies a kind of projective approach, which, in all probability, he developed independently of seemingly cognate attempts like those of Girard Desargues and Philippe de La Hire. His sources can instead be traced back to Nicholas of Cusa for the analogy involving a straight line and an infinite circle and Kepler's *Paralipomena to Witelo* for the transformations of the conics sections into one another and the reference to the point at infinity, which is used in many places (see, e.g., Boscovich 1754a, § 751, b, 61; but there are many other examples in which points at infinity are implicitly or explicitly used). For Cusanus and Kepler as sources of Boscovich, see Guzzardi (2016, 49), where the relevant references are given. Del Centina and Fiocca (2018a) provided a reconstruction of the idea of continuity among curves from Kepler to Boscovich, passing through the projective tradition. Martinović (1986, 172) evoked a possible

"coming back from infinity", corresponding to the hyperbola and the parabola (Boscovich 1754a, §§ 737–742), enabled him to claim this generalization: "In all geometrical figures . . ., if one leg [*crus*] goes away to the infinity, there is always another leg coming back from infinity, either from the same side or from the opposite one, which, at that infinite distance, is connected in a certain manner with the first leg" (ibid., § 743). According to Boscovich, this connection has no other reason than continuity, for "all this is absolutely necessary in order to have the law of continuity very scrupulously preserved [*servata religiosissime*] anywhere in geometry."

An extension to algebra immediately follows. If, in Fig. 6.3, segments such as P_0P_1, P_0P_2, P_0P_3, etc., are considered as magnitudes, then the two regions distinguished by P_0, that have the opposite directions, can respectively be viewed as the negative and the positive domain, and both P_0 and P_∞ can be considered as 'points of transit' from the positive to the negative quantities, and *vice versa*. Boscovich had introduced this argument some paragraphs prior (ibid., § 717); in the subsequent passages, he sought to integrate the geometric "come-back from infinity" and its algebraic interpretation in terms of positive and negative numbers into the common framework of the continuity law. If continuity holds, he claimed, negative magnitudes are those that have gone through the infinite. For him, this signaled a "mystery of the infinite [*mysterium quoddam infiniti*]", since a quantity of this kind "must be said, in a certain manner, to be positive, as well as greater than infinity [*haec dicenda esse quodammodo et positiva, et plusquam infinita*]," even if, in the usual geometry, segments may be considered as magnitudes endowed with signs (ibid., § 753).[7]

Quantities that were greater than infinity were not new to Boscovich's mathematical colleagues, but their status was vague and problematic, to say the least. They had been introduced by John Wallis in a couple of brief passages from his *Arithmetica Infinitorum* (Propositions CI and CIV) that dealt with the quadrature of curves whose equations contain negative exponents. The problem was to find the ratio between the area of a curve and the inscribed rectangle. In an attempt to determine areas of curves of equation $= \frac{1}{x^n}$, where a coordinate is proportional to an inverse power of the other, he had shown that the ratio between the area of the curve and the inscribed rectangle must be expressed by $\frac{1}{n+1}$. Hence, in the hyperbolic curve of equation $= \frac{1}{\sqrt{x}}$, where the exponent $n = -\frac{1}{2}$, the ratio is $1 : -\frac{1}{2} + 1$, or, briefly, $1 : \frac{1}{2}$. For the hyperbola, having equation $y = \frac{1}{x}$ (hence, $n = -1$), the ratio is

mediation of this conception through Bruno's *De la causa, principio et uno*; however, there is no textual evidence for this, and, after all, it hardly seems compatible with Boscovich's education within Jesuit orthodoxy.

[7]Commenting on another construction, given in Fig. 265 of Table VII, Boscovich (1754a, § 753) stated the following: "All this, indeed, pertains to a certain mystery of the infinite and leads to certain analogies; however, in the common geometry [*in Geometria communi*] to the same CM′, negative, corresponds that finite and likewise negative segment CNP[/]." In the figure, CM′ and CNP′ are drawn along a horizontal axis ACB and are on the left side of a vertical axis QCQ′. Magnitudes represented on the right side CB are taken as positive, while the magnitudes represented on the left side (AC) are those "coming back from the infinity" or—in the common interpretation—the negative magnitudes.

$\frac{1}{0}$, which Wallis assumed to be equal to infinity (see, e.g., Prop. XCI). Finally, for curves of equations having $n < -1$, the resulting ratios will be $\frac{1}{-1}, \frac{1}{-2}, \frac{1}{-3}$, etc. But, since the ratio tends to infinity the closer the denominator is to zero and, by assumption, $\frac{1}{0} = \infty$, Wallis concluded that, in curves characterized by $n < -1$, where a positive number is divided by a negative, the ratio between the area of the curve and the inscribed parallelogram will be "greater than infinity: of a kind, that is, that a positive number may be supposed to have to a negative number, or less than zero" (Prop. CIV).[8] However, as Stedall (2004, xxiv) pointed out, the reference to the geometrical operation of the quadrature of curves is essential, since, in other places—and in the *Algebra* above all things—Wallis applied the usual rules of signs (i.e., with 'plus' divided by 'minus' giving 'minus', etc.). As such, his claim of magnitudes 'greater than infinity' is far from having any algebraic significance per se.

This distinction (and the confusion) between the two levels served as the background for later developments. An objection to Wallis' idea came from Pierre Varignon as early as 1697 in a letter addressed to Johann Bernoulli I on August 26, emphasizing the conventional meaning of the mathematical notation, in particular, of operator symbols. According to Varignon, the usual mathematical language deceptively let us "suppose that negative quantities are smaller than zero; it seems to me that this is completely false, since the sign – [minus] only means the subtraction of the thing without changing anything about its value." Expressions like $\frac{a}{-b}$, he claims, do not stand for values "greater than infinity" existing per se; rather "all of these magnitudes greater than infinity [are] nothing but finite magnitudes taken as negative and reversed."[9] Then, Varignon explained this view in a more structured way in an essay that he first sent to Bernoulli (August 20, 1703) and then published with some changes in the *Mémoires de l'Histoire de l'Academie Royale des Sciences* (February 3, 1706):

> The signs + and – are not but signs of operations, that is, addition and subtraction, being accomplished on the magnitudes to which they are applied. They don't modify *per se* a value and are far from being able to make it less than nothing, as should be the case according to

[8]The integral passage in Wallis' Latin text reads as follows: "*Si deniq[ue] ejusmodi Figura ADββ, sic continuo decrescat juxta seriem quae sit reciproca directae indicem habenti unitate majorem; habebit illa ad Parallelogrammum inscriptum rationem plusquam infinitam: qualem nempe habere supponatur numerus posìtivus ad numerum negativum, sive minorem nihilo.*" On Wallis and his concept of magnitudes greater than infinity, see Reiff (1889, 6–8), Scott (1938, 43–46), and Stedall (2004). Maybe surprisingly, such a notion seduced mathematicians well beyond the seventeenth and eighteenth centuries: for a nineteenth-century interpretation, see Peacock (1834, 232–239); for a twentieth-century interpretation, see the often celebrated Waismann (1970, 47–51).

[9]Here, a more extended version of the original text: "*... Sans parler de la difficulté que j'ay toûjours eues sur ce langage qui ne vient que de ce qu'on a suppose que les grandeurs negatives étoient plus petites que zero; ce qui est, ce me semble, tres faux, puisque le signe – ne marque que la soustraction de la chose sans rien changer à sa valeur. Il m'a toûjours semblé que tous ces plus qu'infinis ne sont que des finis négatifs & renversès. Outre qu'a ce compte il faudroit dire que* $\frac{aa}{-b}$ *est moin que rien; ce qui rendroit* les plus qu'infinis *égaux à* des moins que rien." Pierre Varignon à Jean Bernoulli, Paris, le 26 août 1697, in Bernoulli (1988, 126–127; emphasis in the original text).

> Wallis. 1000 *écus* I owe have the same value as 1000 *écus* I would own; 6 to sum up have the same value as 6 to subtract, etc. ... All of the difference lies in the fact that the spaces expressed by these formulae [i.e., where negative denominators occur] are reversed one with respect to the other, as negative expressions mean everywhere in geometry. (Varignon 1706, 17; see also Bernoulli 1992, 101–106)

By Boscovich's time, the idea of quantities greater than infinity had finally turned into a poetic way of speaking. If they had, once upon a time, represented entities endowed with any mathematical reality, their time now seemed to be over. Still, in so far as Boscovich considered an infinite straight line as a circumference of a circle having infinite radius, he appeared to treat Wallis' seminal idea in a serious, as well as realistic, manner: "In the negative magnitudes, one has to replace addition with subtraction, even if they occur because of a point passing through infinity, though the analogous magnitude to that existing before its parting to the infinity is not, absolutely and directly, such a negative magnitude, but that positive magnitude led through infinity, which—according to the above-mentioned idea—is called greater than infinity" (Boscovich 1754a, § 754).[10]

6.3 Mathematical Constraints

In fact, Boscovich (1754a, § 715) pointed out that the passage to negative quantities may occur both through zero and through infinity, though it is essential that "the transit does not happen by jump, but through continuous degrees [*transitus non sit per saltum sed per gradus continuos*]." According to him, the conic sections display "various points of transit through zero and through infinity, and the return from it. By approaching infinity, the points often remain real, either visible in some place or hidden in the infinity, or they may even have become imaginary."[11] In the

[10]Similar expressions can be found in Boscovich's later writings as well. In an addendum to a mathematical work by a former pupil, Francesco Luino (Boscovich 1767, 254, § 39), negative magnitudes are said to be generally ill-defined, so that the commonly widespread "ideas" of their nature are "insecure and confused" ("*incerte ... e confuse*"). Things become even worse, Boscovich maintained, with the "idea of multiplication and division by a negative magnitude, and that of ratio or geometrical proportion, if this expression applies to negative and positive magnitudes taken together." Finally, and at least in a private communication with the mathematician Giuseppe Calandrelli, Boscovich supported a 'relativistic' and 'conventionalist' interpretation of the negative magnitude: "I find that the idea of a multiplication between two negative [numbers] as giving a positive [number] is true only if one considers negative quantities as being smaller than zero; but then again, the ratio between a negative and a positive [number] is not equal to that of this [positive number] and a negative [number]. The fourth proportional is not negative at all, rather it is greater than infinity, according to the idea of the transit through infinity that I have given in the third volume of my elementary mathematics" (Boscovich to Calandrelli, February 11, 1782: Boscovich 2010b, 24).

[11]"*Porro in hujusmodi transformationibus Sectionum Conicarum aliarum in alias habentur punctorum multiplices et transitus per nihilum, ac per infinitum, et regressus inde. Ipsi autem appulsus ad infinitum, vel nihilum saepe puncta retinent in statu reali, vel alicubi conspicua, vel*

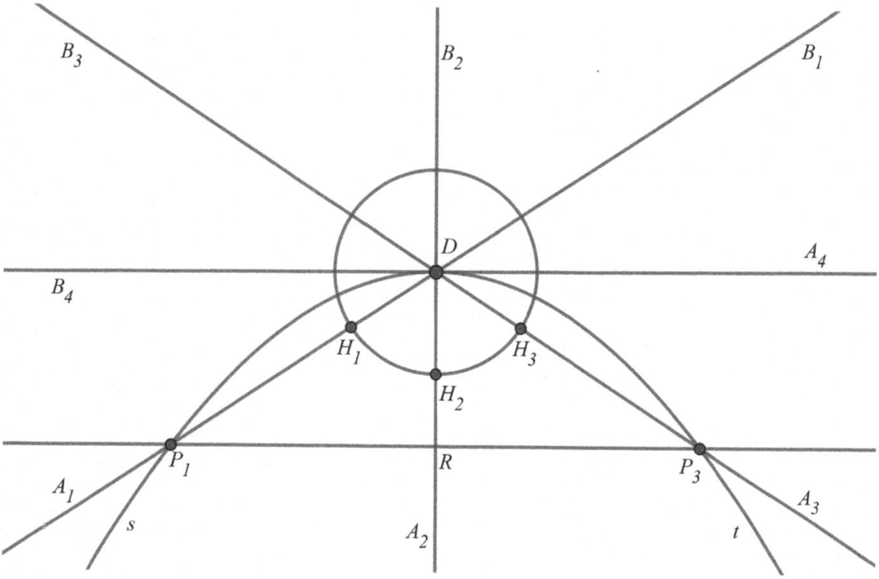

Fig. 6.4 Diagram that corresponds with Boscovich's demonstration that the legs of the parabola join together at a point at infinity (Modern adaptation from Boscovich 1754a, Table VII, Fig. 270)

subsequent paragraphs, he expounded upon numerous examples of such transits. Notably, Del Centina and Fiocca (2018a, 145–154) provided a detailed presentation of Boscovich's method and its underpinnings; here, I focus on the cases of the parabola and the hyperbola "as sort of closed curves" (as they emerge from Boscovich's theory of geometric loci), because they are the crucial cases in light of his natural philosophy.

As shown in our commentary on Fig. 6.3, the transit through infinity is constructed and proved in the case of the straight line. According to Boscovich, we can describe it as the two legs, joined together at the origin of the axis, meeting one another again at a point at infinity. This can also be extended to other curves; of course, parabolas and hyperbolas are more intriguing instantiations of the passage through infinity. To begin with, parabolas do not have rectilinear asymptotes; rather, they infinitely separate from a given straight line. However, as I mentioned in the previous section, Boscovich (1747, § 90) suggested that their legs also join at a point at infinity. His demonstration involved a construction in Boscovich (1754a, §§ 740–741), which reproduced the same argument as in Fig. 6.3. Let parabola *sDt* (see Fig. 6.4), with a vertex in D, be intersected by the straight line B_1DA_1, and let the straight line be rotating on D so that it will reach the positions B_2DA_2, B_3DA_3, etc. Therefore, BDA will meet *sDt* at some point P_n belonging to the leg DS, which

infinito obruta, ibique velut delitescentia, quandoque etiam ad imaginarietatem deturbant" (Boscovich 1754a).

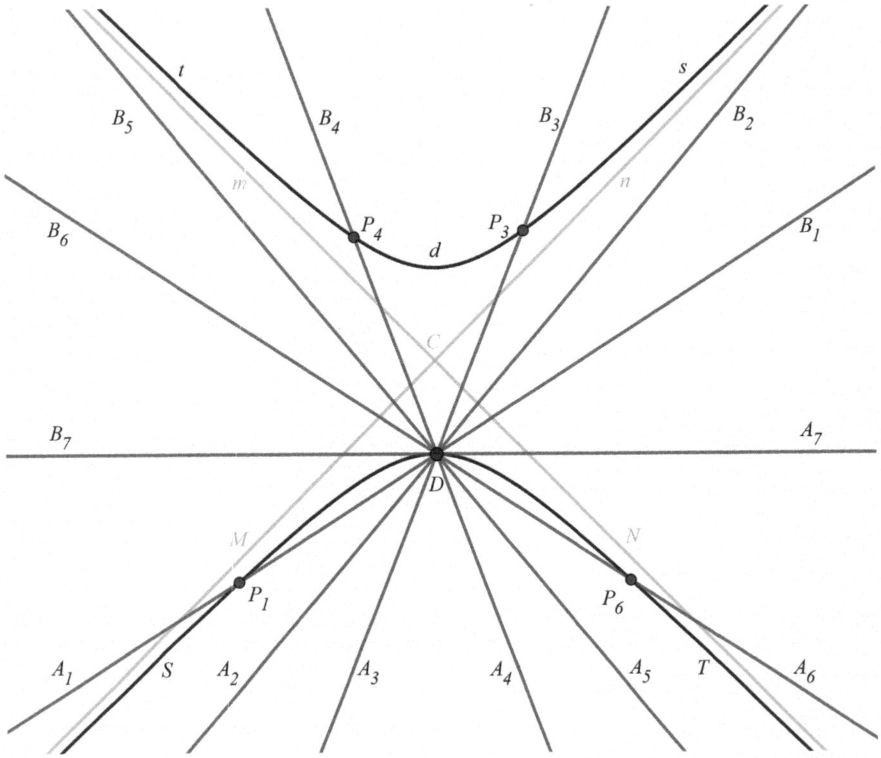

Fig. 6.5 Graph for Boscovich's demonstration that the legs of the hyperbola join together at two points at infinity (Modern adaptation from Boscovich 1754a, Table VII, Fig. 269)

will be further and further from the *x*-axis. Now, if P "recedes beyond all limits along the leg DS, it will also exist" in a determined position, as far as possible. When, after a rotation, B_1DA_1 comes to coincide with the *y*-axis (i.e., it will reach B_2DA_2), then, according to Boscovich, we cannot assign a determined position to P anymore (since it is a point at infinity, and infinity is not a determined quantity).

> However, if we let the straight line advance until [it reaches B_3DA_3], there will immediately be point P_3 belonging to *t*D, and such a point will run along the whole leg until reaching that same point D from which it had departed . . . Therefore DP, as soon as it grows to infinity . . . , will come back to DP_3 having the same direction. On the other hand, the parabola itself will be a continuous curve too, which, in a certain manner, comes back to itself according to the sequence $DP_1s(infinity)tP_3D$. (Boscovich 1754a, § 740)

As for hyperbolas, Boscovich (ibid., § 737) constructed the passage through infinity in an analogous way; however, contrary to parabolas, they do have asymptotes, a fact that has some peculiar consequences. In Fig. 6.5, let a rectangular hyperbola SDT, *sdt*, with asymptotes M*m* and N*n* be intersected by straight line ADB passing through one of its vertexes, such as D. Let ADB be rotating on D; it will intersect the hyperbola at several points P_n so that, as far as A_1DB_1 is concerned,

"P_1 will lie on the leg DS and run along it." When ADB comes to coincide with A_2DB_2, which is assumed to be parallel to the asymptote M*m*, point P "is nowhere, but having travelled through the leg at infinity, it lies in some manner hidden." Let ADB advance and reach A_3DB_3: Now point P re-emerges as P_3 "from an infinite distance in the opposite leg *s*, and P_4 will go along the whole leg *s* by continued motion through A_4DB_4, until it reaches A_5DB_5, parallel to the asymptote N*n*, so that P is nowhere again." However, the motion continues through A_6DB_6 (P_6 being now on the leg DT), and through the tangent A_7DB_7, where P coincides with D. A further explanation follows:

> [W]henever the straight line AB is to have performed half a revolution [i.e., a 180° rotation] by continued motion, in the same manner, the point P will have travelled both branches of the hyperbola by continued motion. The hyperbola itself is to be considered as a certain continuous curve that comes back to itself in a circle [*in orbem redeat*] and joins and connects with itself at those infinite, opposite distances, so that the leg [*d*]*t* is joined with T[D] and [*d*]*s* with S[D]. And then, its continuous path [*ductus*] is DP_1S (*infinity*) sP_3P_4t (*infinity*) TP_6D. (ibid.)

Both the hyperbola and the parabola are characterized by a transit through infinity—but the former passes through it only once, whereas the latter displays a double transit. Boscovich generalized this feature by stating that, in all curves, as a consequence of the law of continuity, which is assumed to be inescapable in geometry, "if a leg goes to infinity, then another leg will always be present, which will come back either from the same side or from the opposite side. This is in some way connected with the former leg at that infinite distance ... For this reason, the legs will always be even in number." He also claimed that, in the curves having asymptotic legs like hyperbolas, there are but four different modes for a leg to come back from infinity. Both asymptotic legs "can come back either from the same side or from the opposite one, and in such a manner that either both legs lie in the same hemiregion [*plaga*] with respect with that asymptote or they lie in the opposite hemiregion."[12] As illustrated in Fig. 6.6 here below, in order to make his point, Boscovich (1754a) drew a set of four diagrams, which he would then summarize in a single diagram in the *Theoria* (see Boscovich 1763, Table I, Fig. 13).

In the Cartesian graph of Fig. 6.6, let us assume the four standard quadrants (I: +,+; II: –,+; III: –,–; IV: +,–). Adopting Boscovich's vocabulary, let us assume that ED is an infinite "leg" of a certain curve, and let us examine the behaviour of ED while approaching the asymptote A′CA, taking for granted that it divides the plane into two hemiregions (*plagae* in Boscovich's jargon)—one including BC and the other including CB′. Finally, let us remember that this serves as an illustration of the passage through infinity. Thus, only positive quantities matter, for the negative will result from such passage as if they were quantities "greater than infinity".

[12] "*Si quod crus in infinitum abeat, semper habebitur crus alterum ex infinito regrediens vel ex eadem parte, vel ex contraria cum ipso in illa infinita distantia connexum quodammodo, quod omnino ad continuitatis legem ubique in Geometria servatam religiosissime est necessarium .... Utrumque crus ... poterit regredi vel ex eadem parte, vel ex opposite, ac vel ita, ut bina crura jaceant respect ejusdem asymptoti ad easdem plagas, vel ita, ut jaceant ad oppositas*" (Boscovich 1754a, § 743).

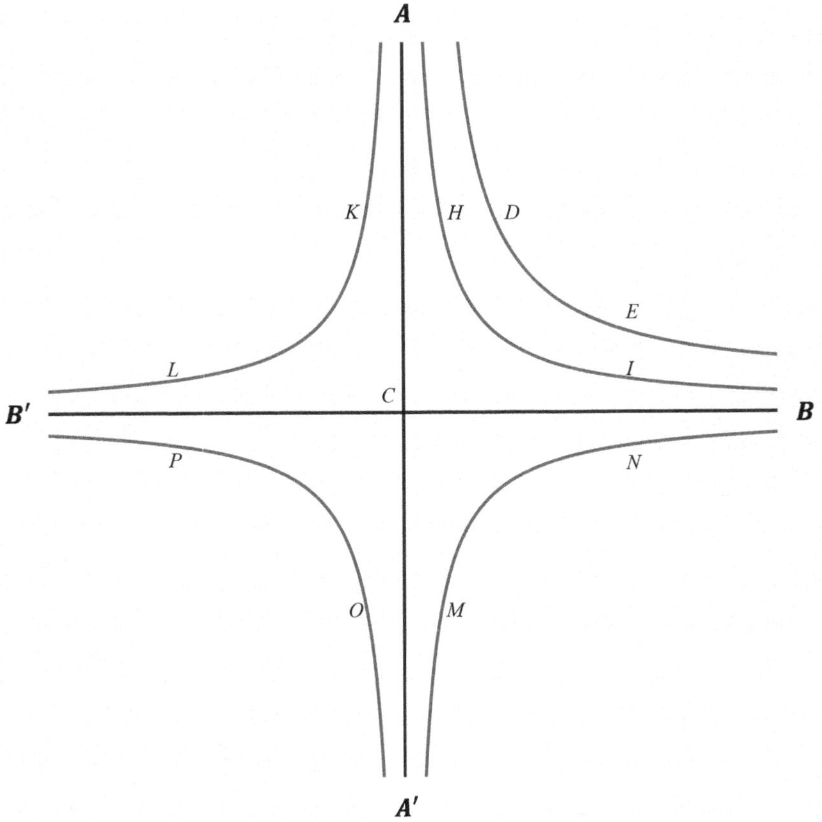

Fig. 6.6 The four modes for an asymptotic leg to come back from infinity. This simplified graph, resuming Figs. 254, 255, 256, 268 of Boscovich (1754a, Table VII), follows the model of the *Theoria* (Boscovich 1763, Table I, Fig. 13), where the same argument is restated

The four cases are generated as follows (I shall present them in accordance with Boscovich 1763, § 163, footnote *i*): In the first case (quadrants I, I), the leg ED goes to infinity following the asymptote CA and comes back as HI on the *same side* of the asymptote and in the *same hemiregion* CB. In the second case (quadrants I, III), ED goes to infinity following CA and comes back as OP on the *opposite side* of the asymptote and in the *opposite hemiregion* CB′. In the third case (quadrants I, IV), ED goes to infinity following CA and comes back as MN; hence, it is on the *opposite side* of the asymptote and in the *same hemiregion* BC. Finally, in the fourth case (quadrants I, II), ED goes to infinity following CA and comes back as KL, which is on the *same side* of the asymptote, but in the *opposite hemiregion* CB′.[13]

[13]Unfortunately, in the Latin-English edition, J.M. Child introduced a new footnote number in § 170 of the *Theoria*, thus splitting Boscovich's note *i* into two parts. So, the four cases are presented in footnote *k* of the Latin-English edition (pp. 132–139), which corresponds to the last two paragraphs of Boscovich's original footnote *i* to § 169. Boscovich (1754a) provided convincing

It might well be that Boscovich had a strong interest in the theory of geometric loci itself. This was partly due to his burdens as the *matheseos professor* at the Roman College and partly due to his reception of the debates over the infinitesimal calculus, from which originated his tormented fascination with infinity (which, of course, traces at least as far back as Boscovich 1741b in *De Natura et usu Infinitorum & Infinite parvorum*). Therefore, as I mentioned in the previous section, it might well be that, in 1747, Boscovich began to develop this (philosophy of) mathematics independently from any interest in the philosophy of nature. But he soon realized that there was a convergence between his manner of thinking about continuity in nature, which had already appeared in the 1745 tract on the living force, and continuity in geometry. Otherwise, he could hardly have reported to his brother that mathematics, and particularly Calculus, should play a crucial role in his ongoing project, the Camaldolese *Ur-Theorie*. Thus, his mathematics might not have been designed for his natural philosophy, but certainly happened to cut across it.

The change in the form of the curve of forces from 1745 (see above, Fig. 6.1a, b) to 1748 (see above, Fig. 6.2) reveals this radicalized mathematical attitude toward natural philosophy. Whereas, in the 1748 graph, the "negative" branch of the curve is drawn, it did not appear in the 1745 forms. If considered in light of Boscovich's mathematical studies by the end of the 1740s, this is not a minor change due to a stubborn will for accuracy, but rather shows his increased awareness of the mathematical constraints of the curve, which could no longer be overlooked.

Let us remember that, as early as 1745, Boscovich had recognized that the curve must be endowed with at least one asymptote, because repulsion must grow asymptotically as the distances are diminished so as to prevent an instantaneous change of velocity—otherwise, the continuity law would be broken. Initially, with Newton's Lemma V in mind, he had termed the curve a kind of parabola; then, by 1748, he realized that, in order to have the desired asymptote, a ratio of two polynomials was needed, hence a curve of a "hyperbolic kind". As assumed in the above-quoted passage from Boscovich (1747, § 90), all curves should preserve continuity so that their infinite legs—including those of a hyperbola—join together at a point at infinity. Therefore, in consideration of the treatise on the transformation of geometric loci (Boscovich 1754a), the "negative branch" of the curve of forces should more properly be interpreted as the leg coming back from infinity. It has no physical meaning at all, as Boscovich (1755a, § 74) recognized while commenting on the "final" form of the curve, presented in his *De lege virium in natura existentium* (ibid.; see Fig. 6.7), for "the distance between points ... can never become

examples of cases two and four through his method of generation of the passage through infinity by the study of the intersection between the curves and a generating straight line. Cases one and three appear more puzzling. According to (ibid., §§ 733–734), the first case involves a cusp at infinity, whereas (ibid., §§ 755–757) the generation of case three involves imaginary quantities (on this, see Del Centina and Fiocca 2018a, esp. 144–149, 161–162). A further exploration of these intriguing aspects exceeds the aims of the present study; I will instead return to cases two and four, as far as they are mentioned and used in the curve of forces.

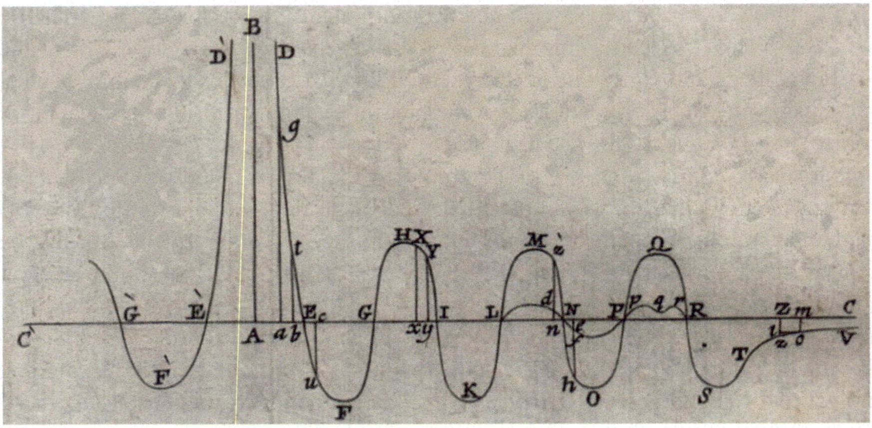

Fig. 6.7 The final form of the curve of forces as it appears in *De lege virium in natura existentium* (Boscovich 1755a). Note that, except for some letters, it is practically identical to the graph plotted in the *Theoria* (Boscovich 1763)

negative." However, it encapsulates and embodies the most important mathematical constraint of the curve: the omnipresent and all-governing continuity.

In other words, Boscovich not only assigned a central role to continuity in mathematics and physics, but the geometrical properties of continuity also model and guide its application within the realm of natural philosophy (on this, see also Schubring 2005, 179–182; Del Centina and Fiocca 2018a, 139). In sum, with his force curve in its final 1755 form, he applied a mathematical feature to natural philosophy that he had explored at length in the treatise on the transformation of geometric loci, studying various ways in which the transit through infinity is possible. Notably, this attitude transpires from his comment accompanying the graph reported above: "*Because of the law of geometrical continuity*, to the asymptotic leg . . . must certainly correspond another leg coming back from infinity, either on the same or on the other side"; moreover, "we shall determine our curve *so that* it will absolutely be of equal form on either side of the asymptote" (emphasis added).[14]

[14]Here is the entire passage from Boscovich (1755a, §74): "The arc D′E′F′G′ should not give us too much trouble, for the distance between the points, because of the first force, and the repulsive area increased at infinity, can never become negative. But, if the leg TV infinitely approaches the inverse square of the distance, a similar and likewise attractive leg must correspond to it on the other side (for the squares of the distances remain positive and, because of the law of geometrical continuity, to the asymptotic leg ED must certainly correspond another leg coming back from infinity, either on the same or on the other side). In view of this, we shall determine our curve so that it will be thoroughly similar and equal on either side of the asymptote AB [*De arcu D′E′F′G′ sito citra asymptotum AB possemus parum esse soliciti, cum distantia punctorum ob primam vim, et aream repulsivam in infinitum auctam, nunquam in negativam abire possit. Sed quoniam si crus TV accedat in infinitum ad rationem reciprocam duplicatam distantiarum, debet ipsi ex parte altera respondere crus simile itidem attractivum (nam distantiarum quadrata positive manent, et cruri asymptotico ED omnino respondere debet ex lege continuitatis geometricae crus aliud ex infinito*

6.4 "Invenire Naturam Curvae"

Of course, Boscovich was aware that every curve can be formulated in algebraic terms, but he tended to prefer a geometrical style, as epitomized in his choosing a phrase such as "the transformation of the geometric loci" as the title to the treatise appended to the third volume of his *Elements of Mathematics*. Algebra and geometry, he claimed, are only modes of expression that refer to the same thing:

> [Natural] philosophers and mathematicians express general connections [*nexus*] among quantities through certain general, indeterminate values, which they display through letters; those expressions are called analytic formulae [*analyticae formulae*], since, in algebra, they are used to signify the common name of analysis. However, mathematicians and philosophers also express and display the same things by means of geometric curves, whose nature they still comprehend [*complectuntur*] by means of analytic formulae.[15]

This applies to the curve of forces as well. Already, Boscovich (1754b, § 169) had noticed that it could "be expressed by means of a very simple algebraic formula [*poterit ea per simplicissimam etiam algebraicam formulam exprimi*]." A year later, he explored this route in depth. First, he connected this attempt with his usual cautionary remarks about forces (determinations):

> In the same manner as God, under the condition of a null distance, can generate a motion in a body and deprive another of it, He can also decide that a reciprocal approach takes place under the condition of a particular distance, whereas a reciprocal separation may take place under the condition of another distance. That is, He can assume, as the condition of mutual approaches and separations, the entire series of distances so that he would assign to those distances such determinations of approaching and separation (which we call, with an arbitrary and customary, but not inappropriate, name, attractive and repulsive forces); and this according to a certain law that we can express by means of an analytic formula or through a geometrical figure.[16]

This expresses a physical constraint that the mathematics of the curve should capture. Namely, it must represent the way in which the determination of points to approach or move away (attraction/repulsion) varies when the distances are varied.

---

regrediens vel ex eadem, vel es opposite parte) ita nostrum determinabimus curvam, ut hinc, et inde ab asymptoto AB sit sui similis penitus et equalis]."

[15] "*Philosophi, ac Mathematici generales quantitatum nexus exprimunt per valores quosdam generales, et indeterminatos, quos literis exhibent, quae quidem expressiones, cum Algebra Analyseos nomine appellari soleat, analyticae formulae dicuntur. Easdem autem et exprimunt, et oculis exhibent per geometricas lineas, quarum tamen Naturam analyticis formulis complectuntur*" (Boscovich 1755a, § 47).

[16] "*Eodem prorsus pacto, quo Deus ex conditione distantiae nullius potest motum progignere in altero corpore, in altero adimere; potest itidem ex conditione tantae distantiae decernere, ut habeatur accessus mutuus, ex conditione autem alterius, mutuus recessus; nimirum assumere pro conditione accessuum, et recessuum totam distantiarum seriem ita, ut determinationes ejusmodi, accessuum mutuorum, vel recessuum (quas vires attractivas, vel repulsivas arbitrario, et jam usitato, nec vero inepto nomine appellamus) secundum certam quondam legem, quam per analyticam formulam exprimere possimus, vel per geometricam figuram oculis subjicere, ipsis distantiis affixerit*" (Boscovich 1755a, § 43).

Yet, there is another important physical/metaphysical constraint: the law of continuity in natural processes. With this constraint, the force (i.e., the determination to move) is infinitely repulsive at infinitely small distances, becomes attractive when distances increase, and may alternatively be turned repulsive and attractive.

The "problem of finding the nature of the curve" ("*Problema: invenire naturam curvae*"), i.e., of finding its analytic formula, can now be set:

> [It is required] that we find the nature of a curve whose abscissae represent distances while the ordinates represent forces that vary as the distances are varied in any manner, and pass from being attractive to being repulsive, and from being repulsive to being attractive, at any number whatsoever of given limit-points. Further, the forces are repulsive at the smallest distances and increase in such a manner that they are capable of annihilating any velocity, however great it may be.[17]

In order to solve the problem, six conditions are expressed in the subsequent paragraph, which are assumed to be necessary and sufficient to build a curve as requested ("*quae requirantur, et sufficient ad habendam curvam, quae quaeritur*": Boscovich 1755a, § 76, 1763, § 118). As we shall see, this is not exactly the case. The six conditions are stated as follows:

1. Let [the curve] be regular and simple, and not compounded of a number of arcs of different curves.
2. Let it cut the axis C′AC of Fig. 6.7 only at certain given points, whose distances, AE′, AE, AG′, AG, and so on, are equal in pairs on each side of A.
3. To a single abscissa, let a single ordinate correspond.
4. If equal abscissae are taken on either side of A, let equal ordinates correspond to them.
5. Let the straight line AB be an asymptote, and let the asymptotic area BAED be infinite.
6. Let the arcs lying between any two intersections vary to any extent, and recede from the axis C′AC to any distances whatsoever, and approximate to any arcs of any curves to any degree of closeness, cutting them, or touching them, or osculating them, at any points and in any manner.[18]

---

[17]"*Invenire naturam curvae, cuius abscissis exprimentibus distantias, ordinatae exprimant vires, mutatis distantiis utcunque mutatas, et in datis quotcunque limitibus transeuntes e repulsivis in attractivas, ac ex attractivis in repulsivas, in minimis autem distantiis repulsivas, et ita crescentes, ut sint pares extinguendae cuicunque velocitati utcunque magnae*" (Boscovich 1755a, § 75; reprised in Boscovich 1763, § 117).

[18]"*Primo: ut sit regularis, ac simplex, et non composita ex aggregato arcuum diversarum curvarum. Secundo: ut secet axem C′AC figurae [8] tantum in punctis quibusdam datis ad binas distantias AE′, AE; AG′, AG; et ita porro aequales hinc, et inde. Tertio: ut singulis abscissis respondeant singulae ordinatae. Quarto: ut sumptis abscissis aequalibus hinc, et inde ab A, respondeant ordinatae aequales. Quinto: ut habeant rectam AB pro asymptoto, area asymptotica BAED existente infinita. Sexto: ut arcus binis quibuscunque intersectionibus terminati possint variari, ut libuerit, et ad quascunque distantias recedere ab axe C′AC, ac accedere ad quoscunque quarumcunque curvarum arcus, quarumcunque curvarum arcus, quantum libuerit, eos secanda, vel tangendo, vel osculando ubicunque, et quomodocunque libuerit*" (Boscovich 1755a, § 76; reprised in Boscovich 1763, § 118).

Finding the "analytical solution" of the problem through such conditions is Boscovich's (1755a) aim. As mentioned above, both the general set-up of the problem and the conditions are reprised in the main text of the *Theoria* (Boscovich 1763, §§ 117–118). On the other hand, the mathematical content of the solution itself, i.e., the majority of 1755a, is included in the *Theoria* as one of its supplements (Boscovich 1758, Suppl. I; Boscovich 1763, Suppl. III; for more details, see Chap. 1, Table 2 of this book). In the following section and subsections, I will comment on each condition.

6.5 Building the Curve

The first condition requires a simple curve, rather than one resulting from a plurality of different curves. In other words, it must be controlled by a single equation expressed as a function of the distances, which is represented on the x-axis. Note that Boscovich assumed "as known the common elements of the ordinary Cartesian algebra, otherwise the thing can in no way be accomplished. So, let the ordinate be called y and the abscissa x." He then specified that, when the curve is under the x-axis, the force is interpreted as attractive; when it is over the x-axis, the force is meant to be repulsive.

As we have seen, since the late 1740s, Boscovich was aware that the curve of forces cannot be a curve of a parabolic kind, but is instead a hyperbolic curve. In other words, it is not a polynomial, but rather a ratio between two polynomials. (Let us call them P and Q, and let them have no common factors.) Moreover, the first arc should represent repulsion growing to infinity as the distance becomes infinitely shorter. To this aim, the inverse of some even power of the distance is needed, so that the curve can be symmetric with respect to the asymptote AB; thus, "let us take $x^2 = z$" (Boscovich 1763, S.III, § 25). On the other hand, let us assume that our curve has any given number of inversions or "cuts" of the x-axis. The number of cuts is not known in advance, since it is a matter of experimental physics. However, so as to allow the curve to pass through them, let us first regard the whole series of possible distances as constants and build the polynomial P with variable $z = x^2$ in the numerator. In accordance with the rules of the polynomial calculation, P should be formed as follows: "Take the values of AE, AG, AI, etc., all with negative sign, and let a be the sum of the squares of all such values, b the sum of the products of all of these squares two at a time, c the sum of the products three at a time, and so on; and let the product of them all together be called f; supposing that the number of these values is m." Hence:

$$P = z^m + az^{m-1} + bz^{m-2} + cz^{m-3} + \ldots + f$$

As Boscovich (1763, S.III, § 25) remarked, "if P is equalled to zero, it is plain that all the roots of this equation will be real and positive, namely, only the squares of the quantities AE, AG, AI, etc., and these will be the values of z. Hence, since $x^2 = z$, and therefore $x = \pm\sqrt{z}$, it is evident that the values of x will be AE, AG, AI, positive, and AE′, AG′, etc., negative".

As for polynomial Q in the denominator, we have already noticed that the ratio $\frac{P}{Q}$, which is not further reducible, must contain the inverse of an even power of the distance in order for the first arc to be repulsive. In other words, Q should be a polynomial with variable z as well as P, and it should contain any number of constants $g, h \ldots l$. Therefore, we obtain the following[19]:

$$Q = z\left(z^r + gz^{r-1} + hz^{r-2} + \ldots + l\right)$$

As Boscovich (1763, S.III, § 27) then concluded, let $P - Qy = 0$ be the difference, as "this equation satisfies all the remaining conditions [1–6] of this curve, and if Q is correctly determined, even the last condition (n. 6) can be satisfied in an infinite number of ways." From this, since $y = \frac{P}{Q}$, the following equation can easily be obtained:

$$y = \frac{z^m + az^{m-1} + bz^{m-2} + cz^{m-3} + \ldots + f}{z(z^r + gz^{r-1} + hz^{r-2} + \ldots + l)}$$

In the above equation, when resuming our data, $z = x^2$, the coefficient a of the numerator is the sum of the squares of the values AE, AG, AI, AL, AN, AP, AR, etc. (i.e., the distances of each limit-point from the origin); b is the sum of the products of such squared values assumed in couples; c is the sum of such products assumed in triplets, and so on; and, finally, f is the product of all of the squared values. The exponent m is the number of the values AE, AG, AI, AL, AN, AP, AR, etc. In the

[19]Boscovich (1763, S.III, § 26) expressed this in a somewhat cryptic manner: "Assume some quantity that is given by z and constants in any manner, so long as it does not have any common measure with P, nor does it vanish when z vanishes; also, if x is made an infinitesimal of the first order, let the quantity become an infinitesimal of the same order, or of a lower order. Such a formula will be any one such as $z^r + gz^{r-1} + hz^{r-2} + \ldots + l$. This, if it is equaled to 0, has any number whatsoever of imaginary roots and any number whatsoever of real roots of whatever kind (provided that none of them will be equal to AE, AG, AI, etc., whether positive or negative), if it is thereafter multiplied by z. Call the product Q. [*Sumatur quaecunque quantitas data per* z, *et constantes quomodocunque, dummodo non habeat ullum divisorem communem cum* P, *ne evanescente* z, *eadem evanescat, ac facta* x *infinitesima ordinis primi, evadat infinitesima ordinis ejusdem, vel inferioris, ut erit quaecunque formula* $z^r + gz^{r-1} + hz^{r-2} + \ldots + l$, *quae posita* = *0 habeat radices quotcunque imaginarias, et quotcunque, et quascunque reales (dummodo earum nulla sit ex iis AE, AG, AI etc., sive positiva, sive negativa), si deinde tota multiplicetur per* z. *Ea dicatur* Q.]"

denominator, the exponent r and the coefficients g, h... l are constants and are chosen so that the ratio is reduced to its lowest terms.

6.5.1 *Simple but Subtle*

The first condition—i.e., the curve must be "regular and simple"—is trivially satisfied, for P and Q have no common factors. As Boscovich (1763, S.III, § 28) explained, if they are "separately equaled to zero, they have no common root ... Hence this equation cannot by division be reduced to two." Therefore, it is not a compound of more equations, so it represents a unique, continuous curve.[20]

It should be noted, however, that Boscovich did not express the equation primarily as a certain function y of x, i.e., in analytical terms (which he derives in the subsequent § 30), but instead adopted the typical language of a geometric locus: $P - Qy = 0$. For the most part, this is perhaps a consequence of Boscovich's overall preference for the geometric approach (and thus it can be viewed as a *longue durée*-effect of his education within the Jesuit mathematical tradition). Yet, there is also a structural reason. Despite the fact that both expressions are extensionally equivalent (they refer to the same curve), there is a subtle but profound intensional difference. Expressing the equation as a geometric locus relates to thinking in terms of a property possessed by objects within a certain domain. In the case at hand, given a(n) (infinite) set of points within space, we simply consider points that have a property defined by $P - Qy = 0$: every point endowed with such a property belongs to that curve. In a certain sense, the points with their property pre-exist the curve to which they belong. On the other hand, expressing a curve according to a function, i.e., $y = f(x)$, relates to posing two different domains in some (possibly univocal) relationship. In our case, we are *searching* for points that stay in a certain relation $R = \frac{P}{Q}$ Let us take any point of the domain "space" (or "distances"); we will only be

[20]According to a widespread view in the mid-eighteenth century, a curve is continuous when expressed by a single equation. This idea of the continuity of a curve is best instantiated by the following passage from Euler (1748, II, chapt. I, §§ 8–9): "Although many different curves can be described mechanically as a continuously moving point, and when this is done, the whole curve can be seen by the eye, we will still consider these curves as having their origin in functions, since they will then be more apt for analytic treatment and more adapted to calculus. Any function [*functio*] of x gives a curve or a straight line, and conversely a curve can define a function. Hence, the nature of each curve is expressed by the function of x .... From this concept of a curve, there immediately follows a division into *continuous* and *discontinuous* or *mixed*. A continuous curve is one such that its nature can be expressed by a single function of x. If the curve is of such a nature that, for its various parts, BM, MD, DM, etc., different functions of x are required for its expression, that is, after one part BM is defined by one function of x, another function is required to express the part MD, then we call such a curve *discontinuous* or *mixed* and *irregular*. This is because such a curve cannot be expressed by one constant law [*legem constantem*], but is formed from several continuous parts." On the continuity of curves in the eighteenth century, see Youschkevitch (1976), Bottazzini (1986, 7–38), Schubring (2005, 25–27), and van Strien (2017, 71–72).

interested in points that satisfy our relation. Here, the relation (the function) in a certain sense pre-exists the points forming the curve. The two aspects are entangled, and both are present in Boscovich's discourse. By giving prominence to the geometrical aspect, he might have aimed to emphasize that the forces are determinations of material points due to their reciprocal distances.

6.5.2 *It Rains Cuts and Dogs*

Boscovich easily demonstrated that the equation $P - Qy = 0$ also satisfies the second condition—i.e., it cuts the x-axis only at certain given points. After all, the equation is designed to that end. Searching for the points at which the curve cuts the axis (briefly, searching for the cuts) relates to looking for solutions in which $y = 0$. In this case, as Boscovich (1763, S.III, § 29) argued, in our equation "when $y = 0$, we have also $Qy = 0$; and therefore, since $P - Qy = 0$, we have $P = 0$." (On the other hand, since $P = 0$, $Q = 0$ is chosen so that they have no common solutions; if $Qy = 0$, then $Q \neq 0$.) Hence, the problem reduces to finding the roots (the solutions) of the equation $P = 0$, and we shall obtain all and only the cuts of the x-axis.

In regard to its physical meaning, this result is highly relevant to Boscovich's discourse, because the cuts represent the limit-points, and thus the points of inversion of the force. From a mathematical point of view, on the other hand, it entails insight, although vague and obscure, into what is usually called the fundamental theorem of algebra, which can be expressed in an intuitive manner by stating that every polynomial with degree n has exactly n solutions (real or complex).[21] In fact, in Boscovich's era, discussions on the theorem were widespread. Early signs of an informal knowledge of it can be traced back to *Arithmetica Philosophica* (1608) by the *Rechenmeister* Peter Roth, who first made the statement (without feeling the need to prove it) that an n-degree equation can have no more than n solutions. Over the course of the seventeenth century, René Descartes, Thomas Harriot and Albert Girard gave more precise formulations, but attempts to prove it failed until Carl Friedrich Gauss.[22] Boscovich's sources about those attempted solutions are not entirely clear, but may reasonably include d'Alembert, as well as Euler (and, of course, many others), whom he might have read before 1755, when *De lege virium in natura existentium* was issued.

---

[21]In modern terms (and following Gauss), the fundamental theorem of algebra can be expressed by saying that, in every algebraic equation of the form $x^m + Ax^{m-1} + Bx^{m-2} + \ldots + M = 0$, or $X = 0$, the polynomial X with real or complex coefficients can be factored into linear factors in the field of complex numbers. (I pick up this definition from van der Waerden 1985, 94; as well known, other equivalent definitions are possible too.) As regards Boscovich, another reference to the theorem, albeit inaccurate, can be seen in his (1755a, § 54): "The higher the degree of a curve is, the higher the number of points at which it can cut a straight line."

[22]On the history of the fundamental theorem of algebra, see Tropfke (1980, 489–496), Struik (1969, 81–102), Gray (1990, 276–279), and Schneider (1993, 59–60).

6.5.3 To Each His Own

Even the third condition,[23] which requires that each abscissa correspond to one and only one ordinate, is effortlessly satisfied. It is in relation to this condition that Boscovich (1763, S.III, § 30) explicitly posed $y = \frac{P}{Q}$, deriving it from the geometric locus $P - Qy = 0$. As he noted, "Hence, for any determinate abscissa x [$\pm x$], there will be a determinate z [x^2]; and thus P and Q will be uniquely determinate. Therefore, y too will be uniquely determinate."

There is a strong connection between this condition and the first one. As mentioned above, once P and Q are chosen so that they do not have common factors, our equation satisfies the "regularity" and "simplicity" of the curve. In fact, the third condition, together with this assumption, embraces the first and also lessens its necessity. Let P and Q have some common factor, so that $P = RP'$ and $Q = RQ'$, where $Q = Q(z)$ is a polynomial in $z = x^2$. Then, the locus of the points belonging to the plane x, y, which satisfies the expression $P - Qy = 0$, is the union of two sets, such as $\Gamma_1 \cup \Gamma_2$, where Γ_1 is the set of pairs (x, y) satisfying the equation $P' - Q'y = 0$. Moreover, Γ_2 is the set of pairs (x, y) satisfying the equation $R = 0$. Since every root of the polynomial R is also a root of P, the roots of R are like $\alpha_1, -\alpha_1 \ldots, \alpha_k, -\alpha_k$, with $k = \deg R(z)$. As a consequence, Γ_2 is formed by the $2k$ straight lines $x = \pm\alpha_1, \ldots, x = \pm\alpha_k$, and all of them are parallel to the y-axis. In this case, adopting Boscovich's terms, the curve determined by $P - Qy = 0$ would not only be neither "simple" nor "continuous", but would also violate the third condition. Specifically, since each $x^* = \pm\alpha_i$, $1 \leq i \leq k$, the equation $P - Qy = 0$ would not univocally determine the ordinate $y^* = \frac{P'}{Q'}$ (under the assumption that at least one of $\alpha_1, -\alpha_1 \ldots, \alpha_k, -\alpha_k$ would not be a root of Q' as well).

We should therefore conclude that the first condition is a consequence of the third, once we assume that the curve is described by an equation of type $P - Qy = 0$, with P and Q being polynomials in x only.

6.5.4 Symmetries

According to the fourth condition, the curve should be symmetrical with regard to the y-axis: for any abscissa, $\pm x$ must correspond to the same ordinate y. This is satisfied because, in the equation $P - Qy = 0$, both P and Q are polynomials in $z = x^2$, which is always positive. Therefore, even if x is taken as negative, we will find the same positive ordinate for each of its values.

The reason why Boscovich posed this condition is not obvious, since his doctrine of the return from infinity allows for another possible symmetry. Let us remember that hyperbolas display a double comeback along both asymptotes. Thus, if the arcs

[23] I am grateful to Claudio Bartocci for the following analysis.

ED, TV are asymptotic, they must certainly come back from infinity. However, ED can return either from the positive part of the y-axis or from the negative part of it. Boscovich considered the former, whereas in the latter case, the arc TV should come back from the other side of the x-axis (i.e., from C′). This possibility is consistent with Boscovich's philosophy of infinity and with all of the conditions except for the fourth. Why, then, was such a condition so important to him? Why did he not consider the other theoretically feasible symmetry?

A possible answer is that he took for granted that the last arc TV reproduces Newton's inverse square law of gravitational attraction, i.e., at substantial (planetary) distances, the force must be proportional to $\frac{1}{x^2}$. Now, let us remember that, in the expression $= \frac{P}{Q}$, the denominator can be expressed as $Q = z\,Q'$, with $z = x^2$ and $Q' = z^r + gz^{r-1} + hz^{r-2} + \ldots + l$. In order to allow that the expression captures the law of gravitational attraction, at sizable distances, Q should (following Boscovich 1755a, § 74) infinitely approximate z, hence $Q' \approx 1$. In other words, the constants $g, h, \ldots, l$ in the denominator must be chosen so that, at long distances, $Q' \approx 1$. In this case, "a similar and likewise attractive leg must correspond to it on the other side" (ibid.) and, as a consequence, the other asymptotic leg is also determined.

It is arguable that this physical requirement (to capture Newton's gravitational law) guided Boscovich in posing $z = x^2$ in the analytical expression of the curve and in letting the whole denominator multiply for z again. On the other hand, this implies that the curve is symmetrical to the y-axis in the manner instantiated by Fig. 6.7.[24]

[24]Boscovich discussed in the *Theoriae vindicatio*, i.e., the last portion of Part I of the *Theoria*, the general significance of the dependence of the force on a potentially complicated function of the distance—in other words, that his law of forces cannot precisely follow Newton's inverse square-law, for "the ordinates [of the curve], since they do not agree with the inverse square of the distances when they remain short, can never follow this relation accurately" (Boscovich 1755a, §72). Maybe stimulated by a reading of the article "Attraction" from the *Encyclopédie* (cited as "*Encyclopaedia Parisiensis*" in Boscovich 1763, S.III, § 61), in the 1763 edition of the *Theoria*, he added seventeen new paragraphs to the analytical solution (see Boscovich 1763, S.III, §§ 60–76), providing a more detailed discussion of a foundational question: Why should the inverse square-law be considered a privileged relation? As he maintained, the appeal to its simplicity is misleading, for the dependence on the square of the distance is as thoroughly complicated as any other relation. It is not true, as he argued further, that an expression such as $\frac{1}{x^2}$ does not include any parameters; it implicitly includes the parameter 1, together with the exponent 2. They may appear simple to our minds, but only depend on the arbitrariness of the Creator. Finally, let us remember that discussions about possible tiny deviations of gravitational attraction from the inverse square law trace back, at least, to Alexis Clairaut's discovery, in 1747, of an alleged anomaly in the motion of the lunar apogee. See Chandler (1975), Waff (1975, 1976), Hankins (1985, 39–40), Calinger (2007, 32–34, 2016, 272–277), Wepster (2010, 9-10).

6.5.5 *Infinitesimals that Cause Infinities*

The fifth condition requires that the y-axis be an asymptote of the curve and that the area BAED be comprised within the asymptote; moreover, the first, repulsive arc is infinite. To ensure that the curve of forces satisfies such a requirement, in the expression $y = \frac{P}{Q}$, let x be infinitely small. Then, according to Boscovich (1763, S. III, 32), "$z = x^2$ will become an infinitesimal of the second order."[25] But the value of P remains finite, because is determined by a final term, f, which does not depend on z, and thus on x; contrarily, Q is entirely dependent on z; hence, it is also infinitely small of the second order. As Boscovich concluded, $y = \frac{P}{Q}$ "will be increased at infinity, so that it becomes an infinity of the second order. Therefore, the curve will have the straight line AB [the y-axis] as an asymptote, and the area BAED will become infinite."

Then, Boscovich continued with an idea that was already present in his 1745 treatise on living forces. Let us imagine two bodies colliding with any velocity, however great. According to him, at a certain very short distance, a repulsive force begins to act so that contact can never occur. In other words, such a repulsive force must be so strong that it annihilates any velocity, however great. Boscovich (1763, § 118, footnote f) expressed this feature in relation to the fifth condition: "This is required because, in mechanics, it is shown that the area of a curve, whose abscissa represent distances and ordinates forces, expresses the increase or decrease of the

[25] Boscovich's classification of infinitesimal into many different orders only partly reflects ours, essentially based on the notion of limit. More precisely, according to Boscovich (1741b, §§ 13–14), "[13] to classify orders, take an arbitrary infinitely small quantity, and call it 'infinitely small of the first order'. With this as a standard, others that depend on its variation can easily be put in determinate classes. Those with a finite ratio to it will also be of the first order. However, if the standard relates to a second quantity as a finite quantity to an infinitely small first order quantity, then the latter will be 'of [the] second order'. Should it happen that, just as a finite quantity relates to a first order infinitesimal, the second order infinitesimal relates to another, this last one will be 'of [the] third order', and so forth .... There are also intermediate orders. A mean proportional between a finite quantity and a first order infinitesimal will be *of intermediate order*, or order $\frac{1}{2}$. Generally, for z, a first order infinitesimal, and a, an arbitrary finite quantity, $\frac{z^m}{a^{m-1}}$ will be a quantity of order m. If $m = 0$, it will be a finite quantity. If m is a positive number, it will be infinitely small and of intermediate order, should m be either a fraction or contain some fraction. Finally, if m is negative, the quantity will be 'infinite of order m,' and, similarly, of an intermediate order if it contains a fraction .... [14] Now, consider a rectangle whose one side is a finite segment, with the other as a first order infinitesimal. It is consistent to call this a 'first order infinitesimal in the genus of surfaces'. The square of a first order infinitesimal will be of a second order. Generally, in arbitrary dimension m, we get an infinitesimal of the first order from a first order infinitesimal in one dimension and a finite quantity of dimension $m - 1$. The mth dimension of a first order infinitesimal quantity will be 'of order m.'" For other details and the historical background of Boscovich's conception of the orders of infinitesimals, see Homann (1993, esp. 414–417). In particular, as Homann demonstrated, these results play a crucial role in the *Theoria*, applied to the subtended areas of the curve, and were used as a "necessary preliminary to the application of the theory to mechanics": see Boscovich (1763, §§ 173–176).

square of the velocity. Hence, in order that the forces should be capable of annihilating any velocity, however great, this area must be greater than any finite area."

Of course, he was aware that, in addition to geometric proofs of the kind provided in *De viribus vivis* with the "geometrization" of the living force compared to momentum (see Sect. 2.3), analytical proofs are also possible. For instance, he referred to them in the *Theoria* (see Boscovich 1763, § 176, footnotes *l*, *m*). In modern terms, given a function $f(x) = \frac{1}{x^n}$, let us consider the interval 0—1 in three cases, where $n = 1$, $n = 2$, and $n = \frac{1}{2}$; we obtain, respectively, $f(x) = \frac{1}{x}$, $f(x) = \frac{1}{x^2}$, and $f(x) = \frac{1}{\sqrt{x}}$. The value of the improper integral $\int_0^1 \frac{1}{x^n}$ is equal to infinity for $f(x) = \frac{1}{x}$ and $f(x) = \frac{1}{x^2}$, whereas it is 2 for $f(x) = \frac{1}{\sqrt{x}}$. This is consistent with the above-quoted statement from Boscovich (1754b, § 164): "In the whole family of the hyperbolas, all those whose ordinates grow less than [the simple ratio to the distance, i.e., $\frac{1}{x}$] have a finite area; those whose ordinates grow more [e.g., $\frac{1}{x^2}$] have an infinite area as well, but infinitely greater than infinity" (Boscovich 1754b, § 164). The phrase "infinitely greater than infinity" ("*infinities magis infinitam*") hints at the distinction between orders of the infinite and the infinitely small quantities that were mentioned above.

6.5.6 Indeterminacy

The statement of the sixth condition appears to be more complicated. It requires that every arc between two "cuts"—e.g., E and G of Fig. 6.7—vary to *any* extent. It can approach the *x*-axis without touching it or endlessly separate from it, provided that the other conditions hold. In order to prove it, Boscovich (1763, S.III, § 33) argued that the expression $y = \frac{P}{Q}$ is satisfied for an infinite number of values of Q, even if we assume only those values that satisfy the first five conditions.[26] Therefore, we can vary any number of arcs between any two intersections of the curve with the *x*-axis; "it follows that they can also be varied, in such a way that the sixth condition is satisfied" (ibid.).

The sixth condition emphasizes that the curve is highly undetermined and shows that the elegant graph of the curve presented in the *Theoria*, which is often viewed as the epitome of Boscovich's theory of natural philosophy, is only one possible curve within an infinite set of possibilities. However, two classes of solution are particularly relevant. Boscovich put them together in Corollary 6 (ibid., § 57). First, "the curve may have [the *x*-axis] C′AC as an asymptote in the directions of C′ and C, in

[26]In other words, Boscovich argued that, if $\Gamma_Q = \{\gamma_1, \gamma_2, \gamma_3, \ldots \gamma_i\}$ is the infinite set of valid solutions of Q and $\Gamma_{Q'} \subset \Gamma_Q$ is the infinite subset of valid solutions of Q, which also satisfy conditions 1–5 (with $\Gamma_{Q'} = \{\gamma'_1, \gamma'_2, \gamma'_3, \ldots \gamma'_i\}$, then every solution γ'_1 of $\Gamma_{Q'}$ is also a solution of Γ_Q. Note that both Γ_Q and $\Gamma_{Q'}$ are taken as infinite because the solutions of Q are in the field of complex numbers (see Boscovich 1763, S.III, § 26). But, in general, not all solutions of Q satisfying condition 6 also satisfy conditions 1–5. (Hence, the sixth condition is independent from the others.)

such a manner that the asymptotic arc is either repulsive or attractive"; second, "any arc intercepted between a pair of limit-points may also go off to infinity and have, for an asymptote, a straight line perpendicular to the axis, however near or far from either limit-point."

According to the former feature, the "last" arc of the curve can approximate (and hence satisfy) Newton's inverse-square law. However, because of the sixth condition, it can approximate and osculate a curve defined by $\frac{1}{x^2}$ for a while and then touch and cut the x-axis at any point, perhaps very far from the origin of the axis (such as in the range of interstellar distances). Therefore, the mathematics of the curve includes both of the possibilities that we have seen in Sect. 4.3, commenting on Boscovich (1753, § 405): God might have created the world according to a universal gravitation that extends at infinity with the same law of the inverse square of the distance, in which case its creation will end in a gravitational catastrophe after a very large but finite number of eons, or the providential God might have created the world in such a way to "avoid this universal ruin" so that, at interstellar distances, some repulsive force rules again. Hence, in such a range of distances, the curve cuts the x-axis again.

As we have seen in the above-quoted Sect. 4.3, this scenario creates the possibility of higher-order aggregates, which is also entailed in the second feature of Corollary 6. By virtue of the sixth condition, which allows for intermediate asymptotes, any arc between two limit-points can separate from the x-axis to any distances, and thus can also go off to infinity and come back from the other side of the axis, along a certain asymptote cutting the x-axis. Boscovich (1755a, §§ 103–110) explored these mathematical features in a series of paragraphs that were not included in Supplement III of the *Theoria*, but are resumed in the main text (Boscovich 1763, Part II, footnote *i* to § 168 and §§ 185–188).

His initial remark in this case reads as follows: "The limit-points discussed so far were obtained through the intersection of the curve with the axis, where the forces vanish at the limit-point. But there may be other limit-points, and the transition from one direction of the forces to another may occur not because of the vanishing of the forces, but through the forces increasing to infinity, i.e., through [intermediate] asymptotic arcs of the curve" (ibid., §185). Let us remember that, as shown in Sect. 6.5.5 in regard to Fig. 6.6, in Boscovich's theory of infinity, there are four modes for an asymptotic leg to return from infinity. As Boscovich contended, "the curve of forces must always go forward," for it describes what happens as distances increase; then, only two cases out of those sketched in Fig. 6.6 are feasible. Either the curve returns to quadrant III (our case 1) so that a repulsive arc is followed by an attractive one, or it comes back in quadrant I (our case 2) so that a repulsive arc is followed by another repulsive arc.[27] According to Boscovich, only case 1 is possible, because the approach of two points belonging to different asymptotic arcs to their common limit-point is restrained by attraction on one side and repulsion on the other side of the asymptote. Otherwise, the aggregates would be instable.

[27]For the sake of precision, there are four possible combinations: quadrant I–II; quadrant IV–II; quadrant IV–III; and quadrant I–III. See Boscovich (1763, §§ 185–186).

However, there is a fundamental difference between the physical-mechanical account in Part II of the *Theoria* and its concise mathematical counterpart in Boscovich (1755a), as well as in the Supplement III to the *Theoria*. This simply means that the mathematics of the curve even encapsulates the hierarchic structure of the Universe, which is essential to Boscovich's theory of matter—it can support and give mathematical form to the insight of a never-ending series of aggregates that globally adhere to a unique law and behave uniformly.

6.6 Epilogue

With the six conditions fixed and the analytical form of the law of forces established, Boscovich's theory of natural philosophy appears complete. The mathematics of the curve encapsulates all of its physical features: the continuity of the processes in nature, for the equation implies that the curve is "regular and simple" (see, Sect. 6.5.1); the limit-points determining the direction of the force, as well as the emerging properties of matter (Sect. 6.5.2); the univocal relationship of distances with force intensities (Sect. 6.5.3); the dependence of attraction on the inverse square of the distance when this is sufficiently great, and thus the consistency with Newton—with gravitation becoming a special case of a generalized law (Sect. 6.5.4); the impossibility of real contact between any two constituents of matter, and thus the discontinuity of material aggregates, since, in the equation solutions, the ordinates increase more "than the simple ratio to the distance" (Sect. 6.5.5); and the plausibility of larger aggregates of aggregates in which our universe would be immersed—as well as the possibility of escaping the cosmological paradox (Sect. 6.5.6).

However, indeterminacy, which Boscovich probably considered the most powerful weapon in the arsenal of his natural philosophy,[28] was extremely weak in the background of the emerging experimental tendencies in the domain of the physical sciences. Indeed, the peculiar analytical form of the law suggests two kinds of investigation for empirical research: first, to determine the limit points—i.e., in our expression $y = \frac{P}{Q}$, to determine the coefficients $a, b, c \ldots f$ in the polynomial P; and second, to search for the constants in Q that serve to determine the amplitude and form of the arcs and the asymptotes. The measurement of the limit-points may appear easier, since Boscovich assumed that *force* means the determination of points

[28]See, for example, Boscovich (1763, § 172). Having explained his idea of a series of different aggregates separated by intermediate asymptotes, he then stated the following: "I only mentioned them here to show how many things there may be well worth considering in that section, and how great is the fertility of this field of investigation, in which possible combinations and possible forms are truly infinitely infinite [*combinationes possibiles, et possibiles formae sunt sane infinities infinitae*]; out of these, those that can be in any way comprehended by the human intelligence are so few compared with the whole that they can be considered as a mere nothing. Yet all of them were seen as present to him in a single act of intuition [*unico intuitu prsesentes vidit*] by *God*, who created the world" (emphasis in the original text).

to approach or separate from each other—so we are only asked to search for the distances at which the one determination vanishes and the opposite one arises. The search for the constants in the denominator seems more complicated, and even comes across as an unrealistic challenge, but one might expect that experimenters could find their way sooner or later.

It is questionable as to whether this could have satisfied Boscovich's colleagues in natural philosophy. Could they have been ready for or even interested in his research program related to finding an ultimate law that could embrace all forces in nature by virtue of an *a priori* method, perhaps grounded on (or at least including as a fundamental part) some exotic mathematics of infinity? As we have seen in Chap. 3, individuals such as Hales and Desaguliers, as well as the generations of Newtonians after them, were primarily preoccupied with measuring *forces*. To do this, they proceeded to a resolute reification of such entities, although under a veil of agnosticism regarding their ultimate nature. They instrumentally used forces in order to gain a wider and more precise physical knowledge, whereas the epistemological preoccupations and the theoretical-mathematical attitude that still characterized Boscovich's approach were gradually but relentlessly abandoned and were ultimately removed from their agenda.[29]

Contrarily, no single original experiment is advanced or described in detail in Boscovich's *Theoria*, nor have there been attempts to solve real problems through equations extrapolated from the theory—two points that are, of course, epistemologically problematic and that look suspect. Furthermore, it seems that any effort to read the *Theoria* as an attempt at quantification falls short of the real expectations of its author. And, after all, Boscovich sometimes referred to his conception as a theory of "general physics" (*physica generalis*), as opposed to the *physica particularis*[30]—

[29]Heilbron (1979, 70–71) effectively described this situation: "Physics ended the century richer in essences than it had begun, and more conscious of its hypothetical character .... This agnosticism and its implied instrumentalism—by all means invoke imponderables, feign hypotheses, multiply forces, if it is necessary to save phenomena conveniently—are characteristic of the Newtonianizing physicists of the second half of the eighteenth century. This was the science of men who grew up familiar with attractions and repulsions and the mathematics needed to treat them; who disposed of more and better data than their predecessors, and lack their epistemological sensibilities." See also Heilbron (1980), and Frängsmyr et al. (1990).

[30]See Boscovich (1757, §§ 1, 39, 76, 90). The most remarkable occurrence is in the 1757 preface to *De materiae divisibilitate*: "I wrote this dissertation already in 1748, when I was asked about the infinite divisibility of matter. It offered me the opportunity to illustrate and extend my theory of general physics [*theoriae Physicae Generalis*], which I had put forth in the 1745 dissertation *De viribus vivis*. I achieved this goal in that same year, 1748, in the dissertation *De lumine*, which was then published. Later, I expounded such a theory in the dissertation *De continuitatis lege*, 1754 ..., as well as in *De lege virium in natura existentium*, 1755 .... That which, in that theory, concerns space and time, I revealed in the Supplements to Stay's philosophy ..." (Boscovich 1757, 131). Boscovich later referred to his pupil Carlo Benvenuti's *Synopsis physicae generalis* (Romae: de Rubeis, 1754) as one of the most trustful presentations of his own conception. Note, however, that Benvenuti's presentation of Boscovich's doctrine raised much deprecation among the superiors of the Order and caused him to be removed from the Roman College. Feingold (2003, 30–31) recounted the episode: "Alessandro Centurione, the Superior of Italy—who would become General

two expressions reflecting the traditional partition of the study of physics within the Society of Jesus. The *physica generalis* was conceived as an investigation and description of the mundane "substances" (in the Aristotelian sense), whereas the *physica particularis* dealt with measurements and experiments on concrete objects.[31] It is possible that Boscovich doubted that any fruitful theory of general physics should confine itself to a commentary of sorts on Aristotle's *Physics*, and it is certainly reasonable to expect that he would have been delighted to see his theory being "confirmed" by some new evidence and not only in (stated) accordance with already known phenomena, as he often claims is the case. And much had changed in the institutional and disciplinary order of Jesuit science since the seventeenth century, when such a partition was the most widespread and unchallenged. Nevertheless, to qualify a work in natural philosophy as a study of "*physica generalis*" was a means to emphasize (and to retrospectively justify against potential criticisms) its speculative traits, as opposed to the measurements and experiments of the *physici particulares*—who, perhaps, should work within the framework that Boscovich had advanced.

In accordance with this unstated expectation, the unnamed *Typographus Venetus*, who authored the preface "To the Reader" of *Theoria*, announced that the book "contains an entirely new system of natural philosophy, which is already commonly called, after its author, the Boscovichean system. Even now, it is publicly communicated in several academies, not only in yearly theses or dissertations, both printed and debated, but also in many handbooks issued for the instruction of the young, in which it is introduced and detailed, and has come to be considered by many as a model."[32] Of course, there is a great deal of commercial rhetoric involved here; yet, to our modern ears, these lines sound absolutely ironical. There was indeed a Boscovich Renaissance in Britain. However, as I have diffusely mentioned in Chaps. 3 and 4, this started in the late eighteenth century and mainly occurred in the nineteenth century as a result of the reification of his concepts of material point and force, which were turned into fundamental entities. To his friend Giovan Stefano

the following year—charged Benvenuti with disobedience and demanded his removal from the College. Significantly, however, Centurione was not particularly troubled by Benvenuti's substitution of Newtonian explanations for those of Aristotle or Descartes. Rather, he complained that the *Synopsis physicae generalis* published by the young Jesuit had effectively turned natural philosophy into a mathematical and experimental science and 'omitted almost entirely the traditional topics of physical ontology, pneumatology, and natural theology'—topics central to the educational objectives of the Society, which sought to unify physics, metaphysics, and theology." Boscovich's intervention, however, mobilized Pope Benedict XIV to, instead of banishing Benvenuti from the College, appoint him to the much less delicate post of professor of Sacred Liturgy.

[31]On the tension between *physica generalis* and *particularis* and their Jesuit practitioners, see Baldini (1992a, 12–11, b, 36–45). See also Heilbron (1979, 98–114, 1993, 2002, 204–216).

[32]"*Typographus Venetus Lectori*", in Boscovich (1763, 2–3).

Conti, however, Boscovich expressed some complaints about the early reception of his theory in France and England.[33]

The problem can be described as generational. The theory of forces, culminating in Boscovich's elegant curve, did not have much to offer a generation of researchers who were approaching the study of nature with a plurality of particular, mainly experimental and quasi-Baconian investigations. Boscovich might have thought that Michell had a theory of (magnetic) force similar to his own, at least superficially. Michell, however, did not present it organically, but provided numerous measurements and a method for building magnets, on the basis of which he suggested, for example, that magnetic attraction and repulsion follow the inverse square-law. Retrospectively, some features of Knight's theory of the two attractive and repulsive principles can appear to be cognate with Boscovich's conception (even if he had not discussed the theory with him). But Knight was primarily an accurate constructor of magnets for compasses. In a time in which most physicists became enthusiastic supporters of a variety of experimental approaches, passionate measurers with scarce care for or interest in the highest speculations of the general physics in Boscovich's style, and opportunist instrumentalists who used the most convenient definition of force under the assumption that forces and their carriers could be measured,[34] it is not surprising that—notwithstanding the rhetoric and concerns of the Venetian printer—the "Boscovichean theory" never became a prevailing subject of scientific debate in the academies, nor was it taken as a standard for the education of the young generations of researchers. Its relative fortune (also in recent time) was due more to a fortunate misunderstanding, throughout the nineteenth century, by many leading English-speaking physicists, rather than being the effect of an uninterrupted influence of its visionary author.[35]

---

[33]The circumstances and Boscovich's impressions about the reception of his natural philosophy are recounted in two of his letters to Conti: those written on April 26, 1760 (see Boscovich 2008, II, 23–24), and February 26, 1762 (ibid., 62). For a commentary, see Guzzardi (2018, 30–33).

[34]See Heilbron (1979, 71–97). Although forty years have passed since his research, I find Heilbron's (1980, 363–364) image of the eighteenth-century "experimental natural philosophy" centered on the use of numbers "for measurement but not for calculation" fresh and insightful. On eighteenth-century experimental culture, see also Frängsmyr et al. (1990), Heilbron (1993), and Home (2003).

[35]On the fortune of Boscovich's *Theoria* and its fluctuations, see Guzzardi (2018, Section 3). Baldini's (2006) detailed study on the reception of Boscovich's *Theoria philosophiae naturalis* shows that only limited circles, mostly connected with the Jesuits and other orders, discussed Boscovich's conception at length in the second half of the eighteenth century. As Baldini noted, "*in continental Europe*, no outstanding scientist devoted a paper to what he considered his main contribution; it was not discussed in any advanced scientific work or taught in any university (a few Jesuit ones excepted); and no academy, including those of which Boscovich was a member, promoted discussion of it" (Baldini 2006, 406, italics in the original text). A couple of reviews appeared shortly after that edition of *Theoria* was first published, respectively, in Germany and France (Mendelssohn 1759; [Montucla] 1760). Yet none of them raised a public debate on Boscovich's work. Note that the French review appeared anonymously in the February *1760* issue of *Journal étranger*, but Boscovich revealed that the author was Étienne Montucla in a letter dated June 4, 1764, to his friend Giovan Stefano Conti (see Boscovich 2008, I, 193). The case of

Such were Boscovich's readers. Some of them might have been impressed by his scholarship and vision, but in order to comprehend his theory, they could not help but reify key terms such as 'force' and 'material points'. Otherwise, they could hardly have had a chance to understand such a powerful mix of (meta)physical speculation and mathematics as that which characterized Boscovich's efforts—in addition to his bold mathematical conjectures about infinity. To comprehend something, they had to misinterpret a great deal. To them, the "single law" that was the brilliant result of the *Theory of natural philosophy* had little to offer: they asked for determining methods and results, it only gave indeterminacy.

Boscovich's reception in Britain (mostly after 1800) was different: see Spencer (1967), Olson (1969), Harmon (1993), and Heilbron (2015).

Concluding Remarks

The Will to Unify, the Force of Plurality

> The contrast between Leibniz and Newton . . . stimulated, throughout the eighteenth century, novel critical efforts everywhere. The reconciliation of the two adversaries now becomes the scientific parole of the age . . . The balance between the demands of thought and those of experience, which had remained out of the reach of mere physical empiricism, now seems feasible only by virtue of a general natural philosophy, which extrapolates its material from immediate observation alone. On the other hand, this should also induce a constructive synthesis of the phenomena and their deduction from a unique fundamental principle. The most important work in the field of natural philosophy of this time, Boscovich's *Theoria philosophiae naturalis* [1763], expresses the distinctive manifestation of such a twofold tendency (Cassirer 1999, II, 426–427).

I am not sure that modern historians and philosophers would endorse this judgement, extracted from Ernst Cassirer's monumental history of philosophy and science. I certainly would not be ready to do so, even if the image of Boscovich that Cassirer drew in this chapter is, in many regards, an illuminating and profound one. (In particular, I think that the "novel critical efforts" of the eighteenth century cannot be easily reduced to any "contrast" between Leibniz and Newton and that, on the other hand, Cassirer greatly underestimates the importance of the eighteenth-century experimental culture.)

In any case, I have not chosen this passage because of its potentially truthful content regarding Boscovich's natural philosophy. Rather, I think that it reveals a great deal, as a matter of contrast, about the approach that I suggested using when reading Boscovich (and perhaps also in doing what Cassirer himself did in many of his works—an integrated history and philosophy of science).

To a reader of the early twentieth century such as Cassirer, who came after Boscovich's "British" Renaissance, his natural philosophy appeared as the expression of a double need for unity. On the one hand, he was the representative of a conceptual claim to unification, derived from the uniqueness of its "fundamental principle". On the other hand, he embodied a historical and quasi-Hegelian tension

L. Guzzardi, *Ruggiero Boscovich's Theory of Natural Philosophy*, Science Networks. Historical Studies 60, https://doi.org/10.1007/978-3-030-52093-9

toward the synthesis between the two belligerent traditions of Newtonianism and Leibnizianism. Did not Boscovich himself (1763, § 1) present his own position as a midpoint between the "systems" of Leibniz and Newton?

Historians, however, should take self-interpretations of this sort as an *explanandum*, not as an *explanans*. They should be viewed as symptoms of a more or less complicated background, rather than as a guiding light endowed with explanatory power. In this, I think that historians and philosophers of science should put Percy Bridgman's (1927, 7) philosophical remark that "the true meaning of a term is to be found by observing what a man does with it, not by what he says about it" to good use. To paraphrase, the most important feature of a theory is discoverable by observing what it does, not by what it or its author claims it does.

In the same vein, throughout this book, I have sought to emphasize certain structural aspects of Boscovich's theory of natural philosophy. Of course, its most important trait is that all physical phenomena are reduced to a single law. This, however, primarily applies to material points and their hierarchical aggregates. Moreover, Boscovich's notion of force, as it emerges from the *Theoria*, as well as from his early treatises on mechanics and natural philosophy (first of all, the 1745 dissertation *De viribus vivis*), is typically Newtonian and very much in debt to the concept of determination. As I have argued, this challenges a noticeable tradition in the studies of Boscovich—that his theory can be understood as a kind of dynamical theory of matter on the one hand and as an anticipation of sorts or a proto-example of a field theory on the other hand. Not only do supporters of this tradition (from Faraday himself to Jammer) ambiguously interpret historical evidence, but the theories are structurally different, no matter what their authors seem to maintain. For example, when reviewing Boscovich's statement that Michell's theory was, in many respects, similar to his own, the prudent historian should adopt a philosophically cautionary stance. The historian will primarily look at what the two theories actually do, compare them on such a basis, and then take the author's words as a symptom of what needs to be explained and investigate the kind of similarities that the author could have envisaged. This also provides the reason why this book does not include a detailed analysis—except for the present critical remarks—of the above-mentioned statement that Boscovich's theory is a midpoint between Newton and Leibniz, a reason that in no way involves a belief on my part that Boscovich does not owe a debt to them in any fundamental way.

I think, in fact, that both Leibniz and Newton are individually important to him, and that we should be careful when distinguishing their respective roles—for example, concerning the concept of force, the matter conception, and the many facets of the continuity law—but I do not see in Boscovich's theory any post-Hegelian synthesis between the unicity of force and the multiplicity of matter points, nor between Leibnizianism and Newtonianism. Instead, I tend to consider his imaginative conception as an episode of what has been called the disunity of science. In each chapter of this book, my effort—and my preoccupation—has been to show the force of the plural traditions lurking behind his natural philosophy, as well as the tensions among the different voices involved. 'Plurality' and 'disunity' do not mean indifference; as such, I have emphasized some aspects that, as far as I can judge on the basis of textual analysis and structural examination, are more important than others. Thus, I have assigned a major role to the mathematical context, which, in turn, includes the Jesuit mathematical tradition in which Boscovich was instructed, his later concerns about the application to physics on the one hand and to Calculus on the other hand, and, finally, the philosophy of the infinite that he developed together with the philosophy of continuity.

In addition, plurality and disunity do not directly imply, but provide a good reason for, openness. With this, I mean that, since I am skeptical about Cassirer's philosophy of history, as well as about any tendency to see any course of events in the long term as a development or, more properly, as an unfolding of fundamental principles that crystallize into epochal turns, I am also skeptical about retrieving a central focus, aim, or theme around which Boscovich's natural philosophy is organized. Therefore, I see no grounds to consider the portrait that I have offered in this book as complete. Rather, I see grounds to regard it as doomed to incompleteness—as with any other attempt, and not only in Boscovich's case. In this special, small domain of Boscovich studies, I see many areas that are only mentioned in the present book and should be investigated more thoroughly: for example, his natural theology, which is partly included in the *Theoria*'s appendix "*Ad metaphysicam pertinens*" on God and the soul, but which permeates many other works as well; a fine-grained description of the mechanical aspects of the theory of natural philosophy; or, again, Boscovich's sources related to elasticity, electricity, magnetism, heat, light, etc., which are present in Part III of the *Theoria* and can possibly be better understood based on his manuscripts and letters. Moreover, I certainly do not see and can hardly imagine many other aspects of Boscovich's theory that may further enrich our image.

Yet, I know that I will not stop being fascinated by every new fold and nuance that our historical-philosophical pencil can continue to add to the portrait of his severe face.

Bibliography[1]

Aguilon F d' (1613) Opticorum Libri Sex philosophis juxta ac mathematicis utiles. Ex officina Plantiniana, apud Viduam et Filios Io. Moreti, Antwerpiae

Alembert JB Le Rond d' (1743) Traité de dynamique. Chez David l'aîné, Libraire, Paris

Alembert JB Le Rond d' (1758) Traité de dynamique. Nouvelle édition, revue et fort augmentée par l'Auteur. Chez David, Libraire, Paris

Allan DGC, Schofield R (1980) Stephen Hales: scientist and philanthropist. Scholar Press, London

Antinori A (2004) Domenico Passionei tra giansenismo e culto dell'antico: il romitorio presso Frascati e la tomba in San Bernardo alle Terme. In: Gambardella A (ed) Ferdinando Sanfelice. Napoli e l'Europa. Edizioni Scientifiche Italiane, Napoli, pp 55–67

Aristotle (1984) The complete works. In: Barnes J (ed) The revised Oxford translation, 2 vols. Princeton University Press, Princeton

Arthur RTW (2014) Leibniz. Polity Press, Cambridge

Arthur RTW (2018) The Labyrinth of continuum. In: Antognazza MR (ed) The Oxford handbook of Leibniz. Oxford University Press, Oxford, pp 275–289

Baldini U (1992a) Boscovich e la tradizione gesuitica in filosofia naturale: continuità e cambiamento. Nuncius 7(2):3–68

Baldini U (1992b) Legem impone subactis. Studi su filosofia e scienza dei Gesuiti in Italia, 1540-1632. Bulzoni, Roma

Baldini U (1996) La formazione scientifica di Giovanni Battista Riccioli. In: Pepe L (ed) Copernico e la questione copernicana in Italia. Olschki, Firenze, pp 123–182

Baldini U (1998) Die Philosophie und die Wissechaften im Jesuitenorden. In: Grundriss der Geschichte der Philosophie, begründet von Friedrich Ueberweg. Die Philosophie des 17. Jahrhundert, Bd. 1/2. Schwabe, Basel, pp 669–769

Baldini U (2003) The Academy of Mathematics of the Collegio Romano from 1553 to 1612. In: Feingold M (ed) Jesuit science and the Republic of Letters. MIT, Cambridge, pp 47–98

[1]Volumes from *Edizione Nazionale delle Opere e della Corrispondenza di Ruggiero Giuseppe Boscovich* (Roma-Milano 2009 ff.) are quoted as *ENo* or *ENc* (for the section "Opere" or "Corrispondenza", respectively), in addition to the corresponding volume number. They are accessible from http://www.brera.inaf.it/edizionenazionaleboscovich/opere_stampa.html and, respectively, http://www.brera.inaf.it/edizionenazionaleboscovich/corrispondenza.html (both checked in August 2020).

L. Guzzardi, *Ruggiero Boscovich's Theory of Natural Philosophy*, Science Networks. Historical Studies 60, https://doi.org/10.1007/978-3-030-52093-9

Baldini U (2006) The reception of a theory: a provisional syllabus of Boscovich literature, 1746-1800. In: O'Malley JW, Bailey GA, Harris SJ, Kennedy TF (eds) The Jesuits II. Cultures, sciences, and the arts, 1540-1773. University of Toronto Press, Toronto, pp 405–450

Baliani GB (1646) De motu naturali gravium solidorum et liquidorum. Ex Typographia Io. Mariae Farroni, Genuae

Barrow JD (2007) New theories of everything. Oxford University Press, Oxford

Bayle P (1740) Dictionnaire historique et critique, 5th edn, 4 vols. [various publishers], Amsterdam

Beeley P (1996) Kontinuität und Mechanismus. Zur Philosophie des jungen Leibniz in ihrem ideengeschichtlichen Kontext. Studia Leibnitiana. Supplementa, 30. Franz Steiner, Stuttgart

Beretta F (2008) L'héliocentrisme à Rome, à la fin du XVIIe siècle: une affaire d'étrangers? Aspects structurels d'un espace intellectuel. In: Romano A (ed) Savants étrangers et cosmopolitisme de la culture scientifique romaine. Collection de l'Ecole française de Rome, Rome, pp 529–554

Bernoulli J (1988) Der Briefwechsel von Johann Bernoulli, bearbeitet und kommentiert von Pierre Costabel und Jeanne Peiffer, Bd. 2, Der Briefwechsel mit Pierre Varignon. Erster Teil. Birkhäuser, Basel

Bernoulli J (1992) Der Briefwechsel von Johann Bernoulli, bearbeitet und kommentiert von Pierre Costabel und Jeanne Peiffer, Bd. 3, Der Briefwechsel mit Pierre Varignon. Zweiter Teil. Birkhäuser, Basel

Bertoloni Meli D (1990) The relativization of centrifugal force. Isis 81(1):23–43

Blair A (2006) Natural Philosophy. In: Park K, Daston L (eds) The Cambridge history of science. Cambridge University Press, Cambridge, pp 363–406

Borgondio O (1714) De telluris motu in orbe annuo ex novis observationibus impugnato theses mathematicae. Komarek, Romae

Borgondio O (1725) De situ telluris exercitatio geographica. Komarek, Romae

Borgondio O (1732) Hypothesis planetarum elliptica Exercitationis Astronomicae gratia explicanda. Komarek, Romae

Bos HJM (1980) Mathematics and rational mechanics. In: Rousseau GS, Porter R (eds) The ferment of knowledge. Studies in the historiography of eighteenth-century science. Cambridge University Press, Cambridge, pp 327–356

Boscovich RG (1735) De Solis ac Lunae defectibus Carmen, manuscript. Ruggero Giuseppe Boscovich Papers. Bancroft Library, Berkeley. BANC MSS 72/238 cz: Carton 1, Folder 1–20, Item 2

Boscovich RG (1737) De Mercurii novissimo infra Solem Transitu Dissertatio. de Rubeis, Romae

Boscovich RG (1739a) De veterum argumentis pro telluris sphaericitate .... de Rubeis, Romae

Boscovich RG (1739b) Dissertatio de Telluris figura habita in Seminario Romano Soc. Jesu ... Augusti Anno MDCCXXXIX. de Rubeis, Romae

Boscovich RG (1740) De motu corporum projectorum in spatio non resistente dissertatio. de Rubeis, Romae

Boscovich RG (1741a) De inaequalitate gravitatis in diversis terrae locis. De Rubeis, Romae

Boscovich RG (1741b) De Natura et usu Infinitorum & Infinite parvorum dissertatio. Komarek, Romae. In: Pepe L (ed) ENo, I, Opere varie [di matematica e geometria]. An English translation with commentary is provided in Homann (1993)

Boscovich RG (1742a) De Annuis Fixarum Aberrationibus dissertatio. Habita in Collegio Romano Societatis Jesu ... Anno MDCCXLII. Mense Augusto, Die 3. Komarek, Romae

Boscovich RG (1742b) De observationibus astronomicis, et quo pertingat earundem certitudo dissertatio. Habita in Seminario Romano Societatis Jesu ... Anno MDCCXLII, mense Augusti, die XXVIII. Typis Antonii de Rubeis, Romae

Boscovich RG (1742c) Disquisitio in universam astronomiam. Publicae disputationi proposita in Collegio Romano Societatis Jesu. Anno 1742. Mense Decembri Die 16. Komarek, Romae

Boscovich RG (1743) De motu corporis attracti in centrum immobile viribus decrescentibus in ratione distantiarum reciproca duplicata in spatiis non resistentibus. Komarek, Romae

Boscovich RG (1744a) Dissertatio de Telluris figura. Typis Antonii de Rubeis, Romae (1739; reprint in Memorie sopra la Fisica e Istoria Naturale di diversi Valentuomini, T. II, Lucca, 159–218)

Boscovich RG (1744b) Nova methodus adhibendi phasium observationes in eclipsibus lunaribus ad exercenda Geometriam, et promovendam Astronomiam. Komarek, Romae (reprint in Memorie sopra la Fisica e Istoria Naturale di diversi Valentuomini, Lucca 1747, tomo III)

Boscovich RG (1745) De viribus vivis. Komarek, Romae (New edition in: Guzzardi L, Bevilacqua F (eds) ENo, VI, Opere precedenti la Theoria)

Boscovich RG (1754a) De transformatione locorum geometricorum, ubi de continuitatis lege, ac de quibusdam Infiniti mysteriis. In: Elementorum universae matheseos tomus III. Continens sectionum conicarum elementa. Salomoni, Romae (New edition in: Pepe L (ed) ENo, II, Elementa Universae Matheseos. Elementi di geometria, T. 3, 709–1203)

Boscovich RG (1754b) De continuitatis lege et ejus consectariis pertinentibus ad prima materiae elementa eorumque vires. Salomoni, Romae (New edition in: Guzzardi L, Bevilacqua F (eds) ENo, VI, Opere precedenti la Theoria)

Boscovich RG (1746) De Cometis dissertatio. Komarek, Romae

Boscovich RG (1747) Dissertatio de maris aestu. Komarek, Romae

Boscovich RG (1748) Dissertationis de lumine, pars secunda. Komarek, Romae

Boscovich RG (1749) Sopra il turbine che la notte tra gli XI, e XII giugno del MDCCXLIX danneggiò una gran parte di Roma. Pagliarini, Roma

Boscovich RG (1755a) De lege virium in natura existentium. Salomoni, Romae (New edition in: Guzzardi L, Bevilacqua F (eds) ENo, VI, Opere precedenti la Theoria)

Boscovich RG (1755b) Supplementa. In: Philosophiae recentioris a Benedicto Stay versibus traditae libri X, cum adnotationibus, et supplementis Rogerii Josephi Boscovich, Tomus I. Palearini, Roma

Boscovich RG (1757) De materiae divisibilitate et principiis corporum. Memorie sopra la fisica IV, pp 131–258 (New edition in: Guzzardi L, Bevilacqua F (eds) ENo, VI, Opere precedenti la Theoria)

Boscovich RG (1758) Philosophiae naturalis theoria redacta ad unicam legem virium in natura existentium. Officina Libraria Kaliwodiana, Prostat Viennae Austriae (New edition in: Guzzardi L (ed) ENo, VII, Theoria philosophiae naturalis)

Boscovich RG (1763) Theoria philosophiae naturalis redacta ad unicam legem virium in natura existentium, Editio veneta prima. Remondini, Venetiis (New edition In: Guzzardi L (ed) ENo, VII, Theoria philosophiae naturalis. English translation quoted after the Latin-English edition: A theory of natural philosophy. Edited by James Mark Child, The Open Court, New York, 1922)

Boscovich RG (1767) Appendice: Su i logaritmi delle quantità negative. In: Pepe L (ed) ENo, I, Opere varie [di matematica e geometria]. Originally published in Luino F, Delle progressioni e serie, libri due. Galeazzi, Milano, pp 239–265

Boscovich RG (2001) De continuitatis lege. Über das Gesetz der Kontinuität. Übersetz und herausgegeben von J. Talanga. Universitätsverlag Winter, Heidelberg

Boscovich RG (2008) Carteggio con Giovan Stefano Conti. In: Proverbio E (ed) ENc, V, in two tomes

Boscovich RG (2010a) Carteggio con Bartolomeo Boscovich. In: Proverbio E, Rigutti M (eds) ENc, II.

Boscovich RG (2010b) Carteggi con corrispondenti diversi. Da Antonio Caccia a Pietro Corer. In: Capecchi D (ed) ENc, IV

Boscovich RG (2012a) Les Eclipses. Poëme en six chants (1779). In: Guzzardi L (ed), ENo, XIII/2

Boscovich RG (2012b) Carteggio con Natale Boscovich. In: Proverbio E (ed) ENc, III, in two volumes

Boscovich RG (2012c) Lettere in croato. In: Kritzman Malev T (ed) ENc, XIII/2

Boscovich RG (2012d) Carmina, poesie, ecloghe, epigrammi. In: Kritzman Malev T (ed) ENo, XIV

Bridgman PW (1927) The logic of modern physics. Macmillan Co., New York

Bottazzini U (1986) The higher calculus: a history of real and complex analysis from Euler to Weierstrass. Springer, New York

Boyer CB (1947) Note on epicycles & the ellipse from Copernicus to Lahire. Isis 38(1/2):54–56

Breger H (1990) Das Kontinuum bei Leibniz. In: Lamarra A (ed) L'infinito in Leibniz. Problemi e terminologia. Edizioni dell'Ateneo, Roma, pp 53–67

Calinger RS (2007) Leonhard Euler: life and thought. In: Breadley R, Sandifer E (eds) Leonhard Euler: life, work and legacy. Elsevier, Amsterdam, pp 5–60

Calinger RS (2016) Leonhard Euler: mathematical genius in the enlightenment. Princeton University Press, Princeton

Caracciolo A (1968) Domenico Passionei tra Roma e la repubblica delle lettere. Edizioni di storia e letteratura, Roma

Carpenter AT (2011) John Theophilus Desaguliers. A natural philosopher, engineer and freemason in Newtonian England. Continuum Publishing, New York

Carter B (1974) Large number coincidences and the anthropic principle in cosmology. In: IAU symposium 63 proceedings, confrontation of cosmological theories with observational data. Reidel, Dordrecht, pp 291–298. (republished in General Relativity and Gravitation, Vol. 43(11), 2011, pp 3225–3233)

Casini P (1966) Carlo Benvenuti. In: Dizionario Biografico degli Italiani, vol 8. Istituto della Enciclopedia Italiana, Roma, pp 661–663

Casini P (1970) Orazio Borgondio. In: Dizionario Biografico degli Italiani, vol 12. Istituto della Enciclopedia Italiana, Roma, pp 777–779

Casini P (1971) Boscovich, Ruggiero Giuseppe. In: Dizionario Biografico degli Italiani, vol 13. Istituto della Enciclopedia Italiana, Roma, pp 221–230

Casini P (1983) Newton e la coscienza europea. il Mulino, Bologna

Cassirer E (1999) Das Erkenntnisproblem in der Philosophie und Wissenschaft der neueren Zeit. Zweiter Band. In: Gesammelte Werke. Hamburger Ausgabe, herausgegeben von B. Recki, Bd. 3. Meiner, Hamburg (First edition 1907, Bruno Cassirer, Berlin

Chandler P (1975) Clairaut's critique of Newtonian attraction: some insights into his philosophy of science. Ann Sci 32(4):369–378

Child JM (1922) Introduction. In: Boscovich RG, A theory of natural philosophy. Edited by James Mark Child, The Open Court, New York, pp xi–xix

Clavius C (1582) Modus quo disciplinae mathematicae in scholis Societatis possent promoveri. In: Lukács L (ed) Monumenta Paedagogica Societatis Iesu. Institutum Historicum Societatis Iesu, Romae, VII, pp 115–117 (1992)

Clavius C (1611a) Operum Mathematicorum Tomus Primus. Complectens Commentaria in Euclidis elementa geometrica.... Hierat, Eltz, Moguntiae

Clavius C (1611b) Operum Mathematicorum Tomus Tertius. Complectens Commentarium in Sphaeram Ioannis de Sacro Bosco.... Hierat, Eltz, Moguntiae

Cleary JJ (1995) Aristotle and Mathematics. Brill, Leiden

Conti GS (1998) Lettere a Ruggiero Giuseppe Boscovich, vol II (1771–1784). A cura di E. Proverbio. Accademia Nazionale delle Scienze detta dei XL, Roma

Costabel P (1960) Leibniz et la dynamique: les textes de 1692. Hermann, Paris

Costabel P (1961) Le De Viribus Vivis de R. Boscovich ou de la Vertu des Querelles de Mot. Arch Int Hist Sci 14:3–21

Courtine J-F (1990) Suárez et le système de la métaphysique. PUF, Paris

Crombie AC (1975) Sources of Galileo's early natural philosophy. In: Bonelli MLR, Shea WR (eds) Reason, experiment, and mysticism in the scientific revolution. Science History, New York, pp 157–175

Čuljak Z (1995) Some aspects of explanation in Boškovič. Int Stud Philos Sci 9(3):73–84

Čuljak Z (1998) Hypothesen und Phänomene. Die Erkenntnis- und Wissenschaftstheorie Ruder Boskovics zwischen Antirealismus und Realismus. Ergon Verlag, Würzburg

Čuljak Z (2008) Einige wissenschaftstheoretische Aspekte von Boškovićs Begründung seiner Theorie der Naturphilosophie: Methodischer Realismus. In: Grössing H, Ullmaier H (eds)

Ruđer Bošković (Boscovich) und sein Modell der Materie. Verlag der Österreichischen Akademie der Wissenschaften, Wien, pp 67–81
D'Agostino S (1989) From Boskovic's notes on the work of Benedict Stay: a criticism of one of Newton's alleged proofs of absolute motion. Memorie della Società Astronomica Italiana 60 (4):759–768
de Gandt F (1995) Force and geometry in Newton's principia. Princeton University Press, Princeton
de Cottignies GF (1667) Pilae motae et quiescentis considerationes Physicomathematicae, manuscript. Roma, Biblioteca Nazionale. Fondo Gesuitico 331
De Pace A (1993) Le matematiche e il mondo. Franco Angeli, Milano
Dear P (1987) Jesuit mathematical science and the reconstitution of experience in the early seventeenth century. Stud Hist Philos Sci 18(2):133–175
Dear P (1995) Discipline & experience. The mathematical way in the scientific revolution. The University of Chicago Press, Chicago
Del Centina A, Fiocca A (2018a) Boscovich's geometrical principle of continuity, and the 'mysteries of the infinity'. Hist Math 45(2):131–175. https://doi.org/10.1016/j.hm.2018.02.002
Del Centina A, Fiocca A (2018b) 'A masterly though neglected work', Boscovich's treatise on conic sections. Arch Hist Exact Sci 72(4):453–495. https://doi.org/10.1007/s00407-018-0213-3
Desaguliers JT (1739) Some thoughts and conjectures concerning the cause of elasticity. Philos Trans 41(1739–1741):175–185 (with one Table)
Desaguliers JT (1742a) A dissertation concerning electricity. W. Innys and T. Longman, London
Desaguliers JT (1742b) Some conjectures concerning electricity, and the rise of vapours. Philos Trans 42(1742–1743):140–143
Desaguliers JT (1744) A dissertation on the cause of the rise of vapours and exhalations in the air. In: A course of experimental philosophy, vol 2. Innys-Senex-Longman, London, pp 336–350
Doyle J (2010) In: Salas VM (ed) Collected studies on Francisco Suárez, S.J. (1548–1617). Leuven University Press, Leuven
du Châtelet É (1742) Institutions de physique. Aux depens de la compagnie, Amsterdam
Du Fay CF d C (1733) A letter from Mons. Du Fay, F. R. S. and of the Royal Academy of Sciences at Paris, to his Grace Charles Duke of Richmond and Lenox, concerning electricity. Philos Trans 38:258–266
Duchesneau F (1994) La dynamique de Leibniz. Vrin, Paris
Ducheyne S (2012) The main business of natural philosophy: Isaac Newton's natural-philosophical methodology. Springer, Dordrecht
Ducheyne S (2014) Newton on action at a distance. J Hist Philos 52(4):675–701
Eschinardi F (1689) Cursus physicomathematicus ... Pars prima. De cosmographia. Tomus primus continens duplicem tractatum. Primum de sphaera. Secundum de astronomia .... Ex Typographia Ioannis Iacobi Komarek, Rome
Euler L (1736) Mechanica sive motus scientia analyticae exposita, vol 1. Edidit Paul Stäckel. In: Leonhardi Euleri Opera Omnia, Series II, Opera mechanica et astronomica, vols 1–2. Teubner, Leipzig (1912)
Euler L (1748) Introductio in Analysin Infinitorum, 2 vols. Bousquet, Lausannae (Eng. Transl. by J. Blanton, Introduction to the Analyses of the Infinite, 2 vols. New York: Springer 1990)
Faraday M (1844) A speculation touching electric conduction and the nature of matter. In: Experimental researches in electricity II. Richard and John Edward Taylor, London, pp 284–293
Faraday M (1855) On some points of magnetic philosophy. In: Experimental researches in electricity III. Richard and John Edward Taylor, London, pp 566–574
Feingold M (1993) A Jesuit among Protestants: Boscovich in England c. 1745-1820. In: Bursill-Hall P (ed), R.J. Boscovich:Vita e attività scientifica – his life and scientific work. Istituto della Enciclopedia Italiana, Roma, pp 511–526
Feingold M (2003) Jesuits: Savants. In: Feingold M (ed) Jesuit science and the Republic of letters. MIT, Cambridge, pp 1–45
Feldhay R (1987) Knowledge and salvation in Jesuit culture. Sci Context 1(2):195–213

Feldhay R (1998) The use and abuse of mathematical entities: Galileo and the Jesuits revisited. In: Machamer P (ed) The Cambridge companion to Galileo. Cambridge University Press, Cambridge, pp 80–146

Feldhay R (1999) The cultural field of Jesuit science. In: O'Malley JW, Bailey GA, Harris SJ, Kennedy TF (eds) The Jesuits: culture, learning and the arts, 1540–1773. University of Toronto Press, Toronto, pp 107–130

Fernandes E (2013) A modern approximation to Pythagoreanism: Boscovich's 'point atomism'. In: Cornelli G, McKirahan R, Constantinos M (eds) On Pythagoreanism. De Gruyter, Berlin, pp 435–482

Festa E (1992) Quelques aspects de la controverse sur les indivisibles. In: Bucciantini M, Torrini M (eds) Geometry and atomism in the Galilean school. Olschki, Florence, pp 193–207

Festa E (1999) Sur l'atomisme dans l'école galiléenne. In: Salem J (ed) L'Atomisme aux XVIIe etXVIII^{e} siècles. Publications de la Sorbonne, Paris, pp 101–117

Fichant M (2016) Les dualités de la dynamique leibnizienne. Lexicon philosophicum. Int J Hist Texts Ideas 4:11–41

Finocchiaro M (ed) (1989) The Galileo affair: a documentary history. University of California Press, Berkeley

Finocchiaro M (2005) Retrying Galileo, 1633-1992. University of California Press, Berkeley

Fisher J (2010) Conjectures and reputations: the composition and reception of James Bradley's paper on the aberration of light with some reference to a third unpublished version. Br J Hist Sci 43(1):19–48. https://doi.org/10.1017/S0007087409990379

Foligno A (2018) The point mass as a model for epistemic representation. A historical & epistemological approach. Ph.D. Thesis. Discussed on 8 March 2018 at the University of Urbino

Frängsmyr T, Heilbron JL, Rider RE (eds) (1990) The quantifying spirit in the eighteenth century. University of California Press, Berkeley

Friedman M (2004) Introduction to I. Kant, metaphysical foundations of natural science. Cambridge University Press, Cambridge, pp vii–xxx

Friedman M (2013) Kant's construction of nature. A reading of the metaphysical foundations of natural science. Cambridge University Press, Cambridge

Gabbey A (1971) Force and inertia in seventeenth-century dynamics. Stud Hist Philos Sci 2 (1):1–67

Gale G (1970) The physical theory of Leibniz. Studia Leibnitiana 2(2):114–127

Gale G (1973) Leibniz's dynamical metaphysics and the origins of the Vis Viva controversy. Systematics 11:181–207

Gale G (1974) Leibniz and some aspects of field dynamics. Studia Leibnitiana 6(1):28–48

Gale G (1988) The concept of 'force' and its role in the genesis of Leibniz's dynamical viewpoint. J Hist Philos 26(1):45–67

Garber D (1985) Leibniz and the foundations of physics: the Middle Years. In: Okruhlik K, Brown JR (eds) The natural philosophy of Leibniz. Reidel, Dordrecht, pp 27–130

Garber D (1992) Descartes' metaphysical physics. Chicago University Press, Chicago

Garber D (2009) Leibniz: body, substance, monad. Oxford University Press, Oxford

Gaukroger S (1982) The metaphysics of impenetrability: Euler's conception of force. Br J Hist Sci 15:132–154

Giacobbe GC (1972a) Il Commentarium de certitudine mathematicarum disciplinarum di Alessandro Piccolomini. Physis 14(2):162–193

Giacobbe GC (1972b) Francesco Barozzi e la Quaestio de certitudine mathematicarum. Physis 14 (4):357–374

Giacobbe GC (1973) La riflessione metamatematica di Pietro Catena. Physis 15(2):178–196

Giacobbe GC (1976) Epigoni nel Seicento della Quaestio de certitudine mathematicarum: Giuseppe Biancani. Physis 18(1):5–40

Giacobbe GC (1977) Un gesuita progressista nella Quaestio de certitudine mathematicarum rinascimentale: Benito Pereyra. Physis 19:51–86

Giusti Doran B (1975) Origins and consolidation of field theory in nineteenth-century Britain: from the mechanical to the electromagnetic view of nature. Hist Stud Phys Sci 6:133–260. https://doi.org/10.2307/27757342

Gómez Camacho F (2004) Espacio y tiempo en la escuela de Salamanca: El tratado de J. de Lugo S.J. "Sobre la composición del continuo". Ediciones Universidad de Salamanca, Salamanca

Granger G-G (1968) Essai d'une philosophie du style. Odile Jacob, Paris (first edition Paris: Armand Colin 1968)

Gray J (1990) Algebra in der Geometrie von Newton bis Plücker. In: von Scholz E (ed) Geschichte der algebra. Wissenschaftsverlag, Mannheim

Grmek M (1996) La méthodologie de Boscovich. Rev Hist Sci 49(4):379–400

Grössing H, Ullmaier H (eds) (2009) Ruđer Bošković (Boscovich) und sein Modell der Materie. Verlag der Österreichischen Akademie der Wissenschaften, Wien

Gualandi A, Bònoli F (2004) Eustachio Manfredi e la prima conferma osservativa della teoria dell'aberrazione annua della luce. In: Leone M, Paoletti A, Robotti N (eds) Atti del XXII Congresso nazionale di Storia della fisica e dell'astronomia. Napoli, IISF, pp 476–481

Gualandi A, Bònoli F (2009) The search for Stellar Parallaxes and the discovery of the aberration of light: the observational proofs of the earth's revolution, Eustachio Manfredi, and the 'Bologna Case'. J Hist Astron 40(2):155–172. https://doi.org/10.1177/002182860904000202

Guerlac H (1977) The continental reputation of Stephen Hales [originally 1951]. In: Essays and papers in the history of modern science. Johns Hopkins University Press, Baltimore, pp 275–284

Gueroult M [1934] (1967) Dynamique et métaphysique leibniziennes. Les Belle Lettres, Paris (Reprinted as Leibniz. Dynamique et Métaphysique. Aubier-Montaigne, Paris)

Guicciardini N (1996) Stars and gravitation in eighteenth century Newtonian astronomy. In: Pepe L (ed) Copernico e la questione copernicana in Italia. Olschki, Firenze, pp 263–280

Guicciardini N (1999) Reading the principia: the debate on Newton's mathematical methods for natural philosophy from 1687 to 1736. Cambridge University Press, Cambridge

Guicciardini N (2009) Isaac Newton on mathematical certainty and method. MIT, Cambridge

Guicciardini N (2011) Newton. Carocci, Roma

Guzzardi L (2015) Personaggi. Ruggiero Giuseppe Boscovich. In: Mantovani D (ed) Almum Studium Papiense. Storia dell'Università di Pavia, vol II. Monduzzi Editoriale Cisalpino, Milano, pp 349–352

Guzzardi L (2016) Introduzione. In: Guzzardi L (ed) ENo, VI, Opere precedenti la Theoria, pp 3–62

Guzzardi L (2018) Introduzione. In: Guzzardi L (ed) ENo, VII, Theoria philosophiae naturalis, pp 13–55

Hacking I (1985) Why motion is only a well-founded phenomenon. In: Okruhlik K, Brown JR (eds) The natural philosophy of Leibniz. Reidel, Dordrecht, pp 131–150

Hacking I (1992) Style for historians and philosophers. Stud Hist Philos Sci 23(1):1–20

Hacking I (2009) Scientific reason. National Taiwan University Press, Taipei

Hales S (1727) Vegetable staticks: or, an account of some statical experiments on the sap in vegetables. W. Innys and R. Manby, and T. Woodward, London

Hall AR (1993) All was light: an introduction to Newton's opticks. Clarendon Press, Oxford

Hankins TL (1985) Science and the enlightenment. Cambridge University Press, Cambridge

Harmon PM (1993) Boscovich and British natural philosophy. In: Bursill-Hall P (ed) R.J. Boscovich: Vita e attività scientifica – his life and scientific work. Istituto della Enciclopedia Italiana, Roma, pp 561–575

Haskell Y (2003) Loyola's Bees. Ideology and industry in Jesuit Latin Didactic Poetry. Oxford University Press, Oxford

Heider D (2014) Universals in second Scholasticism. John Benjamins, Amsterdam

Heider D (2015) Suárez on the metaphysics and epistemology of universals. In: Salas VM, Fastiggi RL (eds) A companion to Francisco Suárez. Brill, Leiden

Heilbron JL (1979) Electricity in the seventeenth and eighteenth centuries. A study of early modern physics. University of California Press, Berkeley

Heilbron JL (1980) Experimental natural philosophy. In: Rousseau GS, Porter R (eds) The ferment of knowledge. Studies in the historiography of eighteenth-century science. Cambridge University Press, Cambridge, pp 357–388. https://doi.org/10.1017/CBO9780511572982.010

Heilbron JL (1993) Weighing imponderables and other quantitative science around 1800. Hist Stud Phys Biol Sci 24(1):1–337. https://doi.org/10.2307/27757720

Heilbron JL (2002) La fisica matematica. In: Storia della scienza, vol VI: L'Età dei Lumi, coordinamento scientifico di J.L. Heilbron. Istituto della Enciclopedia Italiana, Roma, pp 199–216

Heilbron JL (2015) Boscovich in Britain. In: Arabatzis T, Renn J, Simões A (eds) Relocating the history of science. Springer, Cham, pp 99–116

Heilbron JL (2018) From the Roman College to the Royal Society. In: Bevilacqua F, Contardini (ed) Boscovich and his times. Contributions of the Pavia 2011 International Conference. University of Pavia Press, Pavia

Heimann PM (1971) Faraday's theories of matter and electricity. Br J Hist Sci 5:235–257

Heimann PM (1978) Voluntarism and immanence: conceptions of nature in eighteenth-century thought. J Hist Ideas 39(2):271–283

Heimann PM, Mc Guire JE (1971) Newtonian forces and lockean powers: concepts of matter in eighteenth-century thought. Hist Stud Phys Sci 3:233–306. https://doi.org/10.2307/27757320

Hertz HR (1999). Die Constitution der Materie. Eine Vorlesung über die Grundlagen der Physik aus dem Jahre 1884. Herausgegeben von A. Fölsing. Springer, Berlin

Hesse M (1961) Forces and fields. Thomas Nelson & Sons, London

Hetherington NS (1988) Science and objectivity: episodes in the history of astronomy. Iowa State University Press, Ames

Hill E (1961) Roger Boscovich. A biographical essay. In: Whyte LL (ed) Roger Joseph Boscovich. Studies of his life and work on the 20th anniversary of his birth. Allen & Unwin, London, pp 17–101

Hirshfeld AW (2001) Parallax: The race to measure the Cosmos. Freeman, New York

Homann FA (1993) On Boscovich's De natura et usu infinitorum and other mathematical works: Translation and Commentary. In: Bursill-Hall P (ed) R.J. Boscovich. Vita e attività scientifica – his life and scientific work. Istituto della Enciclopedia Italiana, Roma, pp 407–436

Home RW (2003) Mechanics and experimental physics. In: Porter R (ed) The Cambridge history of science, Eighteenth-century science, vol 4. Cambridge University Press, Cambridge, pp 354–374. https://doi.org/10.1017/CHOL9780521572439.016

Hoskin M (1982) Stellar astronomy: historical studies. Science History Publication, Chalfont St Giles

Hoskin M (ed) (1997) The Cambridge illustrated history of astronomy. Cambridge University Press, Cambridge

Iltis C (1967) The Vis Viva controversy: Leibniz to D'Alembert. PhD dissertation. University of Wisconsin

Iltis C (1970a) D'Alembert and the Vis Viva controversy. Stud Hist Philos Sci 1(2):135–144

Iltis C (1970b) Leibniz' concept of force: physics and metaphysics. Studia Leibnitiana 13:143–149

Iltis C (1971) Leibniz and the Vis Viva controversy. Isis 62:21–35

Indorato L, Nastasi P (1993) Boscovich and the Vis Viva controversy. In: Bursill-Hall P (ed) R.J. Boscovich: Vita e attività scientifica – his life and scientific work. Istituto della Enciclopedia Italiana, Roma, pp 169–194

Jammer M (1957) Concepts of force: a study in the foundation of dynamics. Harvard University Press, Cambridge

Janiak A (2007) Newton and the reality of force. J Hist Philos 45(1):127–147

Janiak A (2008) Newton as philosopher. Cambridge University Press, Cambridge

Jardine N (1988) The epistemology of the sciences. In: Schmitt CB, Skinner QRD, Kessler E (eds) The Cambridge history of renaissance philosophy. Cambridge University Press, Cambridge, pp 685–711

Joy L (2008) Scientific explanation from formal causes to laws of nature. In: Park K, Daston L (eds) The Cambridge history of science, Early modern science, vol 3. Cambridge University Press, Cambridge, pp 70–104

Kant I (1992) The Jäsche Logik (1800). In: Lectures on logic (trans, ed: Young JM). Cambridge University Press, Cambridge

Knebel SK (2000) Wille, Würfel und Wahrscheinlichkeit: Das System der moralischen Notwendigkeit in der Jesuitenscholastik 1550-1700. Meiner, Hamburg

Knebel SK (2011) Suarezismus: Erkenntnistheoretisches aus dem Nachlass des Jesuitengenerals Tirso González de Santalla (1624–1705). B. R. Grüner, Amsterdam

Knight G (1754) An Attempt to demonstrate, that all the phænomena in nature may be explained by two simple active principles, attraction and repulsion. J. Nourse, London

Knowles Middleton WE (1975) Science in Rome, 1675-1700, and the Accademia Fisicomatematica of G.G. Ciampini. Br J Hist Sci 8:138–154

Knudsen O, Pedersen KM (1969) The link between 'determination' and conservation of motion in Descartes' dynamics. Centaurus 13:183–186

Kragh H (2009) Contemporary history of cosmology and the controversy over the multiverse. Ann Sci 66(4):529–551. https://doi.org/10.1080/00033790903047725

Lattis J (1994) Between Copernicus and Galileo: Christoph Clavius and the collapse of ptolemaic cosmology. University of Chicago Press, Chicago

Leibniz GW (1673) De vera methodo philosophiae et theologiae ac de natura corporis. In: Schepers H, Schneiders W, Kabitz W (eds) Sämtliche Schriften und Briefe, Band 3 (1672-1676). Akademie Verlag, Berlin, pp 154–159. 1980

Leibniz GW (1686) Brevis demonstratio erroris memorabilis Cartesii ... [Acta eruditorum, 1686]. In: Leibnizens mathematische Schriften, herausgegeben von C.I. Gerhardt, vol VI, Schmidt, Halle 1860, pp 117–123 (English trans in: Loemker LE (ed) Philosophical papers and letters. Springer, Dordrecht, 1969, pp 296–302)

Leibniz GW (1687) Lettre de M.L. sur un principe general utile à l'explication des loix de la nature par la consideration de la sagesse divine, pour servir de replique à la reponse du R. Malebranche [Bayle's Nouvelles de la République de Lettres di Pierre Bayle, July 1687, art. VIII, 744-753]. In: Die philosophischen Schriften, herausgegeben von C.I. Gerhardt, vol. III, Weidmann, Berlin, pp 51–55 (1887) (English trans in: Loemker LE (ed) Philosophical papers and letters. Springer, Dordrecht, 1969, pp 351–356)

Leibniz GW (1695) "Specimen Dynamicum ..." [Part I published in Acta eruditorum, 1695; Part II posthumous]. In: Gerhardt CI (ed) Leibnizens mathematische Schriften, vol VI, pp 234–254. Schmidt, Halle, 1860 (English trans in: Loemker LE (ed) Philosophical papers and letters. Springer, Dordrecht, 1969, pp 118–138)

Leibniz GW (1994) In: Fichant M (ed) La réforme de la dynamique. Vrin, Paris

Leibniz GW (2001) The Labyrinth of the continuum: writings on the continuum problem, 1672-1686. Yale University Press, New Haven

Lützen J (2005) Mechanistic images in geometric form: Heinrich Hertz's principles of mechanics. Oxford University Press, Oxford

Mach E (1919) The science of mechanics. A critical and historical account of its development, 4th edn. The Open Court, Chicago

Malara I (2020) Gravitas, renaissance concept of. In: Sgarbi M (ed) Encyclopedia of renaissance philosophy. Springer, Cham

Manara CF, Spoglianti M (1979) R.G. Boscovich e i precursori di V. Poncelet. Rendiconti del Semininario Matematico di Brescia 3:142–179

Mancosu P (1996) Philosophy of mathematics and mathematical practice in the seventeenth century. Oxford University Press, Oxford

Mandelbaum M (1964) Philosophy, science, and sense perception. The Johns Hopkins Press, Baltimore

Marković Ž (1961) Boscovich's Theoria. In: Whyte LL, Boscovich RJ (eds) Studies of his life and work on the 20th anniversary of his birth. Allen & Unwin, London, pp 127–152

Marković Ž (1968) Ruđe Bošković, 2 vols. Jugoslavenska Akademija Znanosti i Umjetnosti, Zagreb
Marković Ž (1973) Bošković, Rudjer J. In: Gillispie CC (ed) Dictionary of scientific biography. Charles Scribner's Sons, New York, vol II, pp 326–332
Martinović I (1986) Boškovićev prijepor o jednostavnosti pravca iz 1747. godine: izrečeni i prešućeni argumenti. Vrela i prinosi 16:167–179
Martinović I (1987) The fundamental deductive chain of Bošković's natural philosophy. In: Pozaić V (ed) The philosophy of science of Ruđer Bošković. FTI, Zagreb, pp 65–99
Martinović I (1990) Theories and inter-theory relations in Bošković. Int Stud Philos Sci 4:247–262
Martinović I (1991) Bošković's theory of the transformations of geometric loci: program, axiomatics, sources. In: Proceedings of the International Symposium on Ruder Bošković, Dubrovnik, 5th–7th October 1987. Jugoslavenska akademija znanosti i umenosti, Zagreb, pp 79–86
Martinović I (1993) Boscovich on the problem of Generatio Velocitatis: genesis and methodological implications. In: Bursill-Hall P (ed) R.J. Boscovich: Vita e attività scientifica – his life and scientific work. Istituto della Enciclopedia Italiana, Roma, pp 59–79
Martinović I (2015) The concept of the Infiniti mysteria in Bošković's geometrical investigations. Prilozi za istraživanje hrvatske filozofske baštine 41(81):61–90
Maxwell JC (1865) A dynamical theory of the electromagnetic field. Philos Trans 155(1865):459–512
Mayaud P-N (1997) La condamnation des livres coperniciens et sa révocation à la lumière de documents inédits des Congrégations de l'Index et de l'Inquisition. Editrice Pontificia Università Gregoriana, Roma
Mazzuchelli GM (1762) Gli Scrittori d'Italia, II/3. Bossini, Brescia
McCormmach R (2012) Weighing the World: the Reverend John Michell of Thornhill. Springer, Dordrecht
McGuire JE (1970) Atoms and the 'analogy of nature': Newton's third rule of philosophizing. Stud Hist Philos Sci A 1(1):3–58. https://doi.org/10.1016/0039-3681(70)90024-5
McGuire JE (1996) Force, active principles, and Newton's invisible realm. In: Tradition and innovation. Newton's Metaphysics of Nature, Kluwer, Dordrecht et a., pp 193–194
McMullin E (1978) Newton on matter and activity. University of Notre Dame Press, Notre Dame
McMullin E (2001) The impact of Newton's principia on the philosophy of science. Philos Sci 68 (3):279–310
Mendelssohn M (1759) Briefe, die neueste Literatur betreffend. Friedrich Nicolai. I. Theil, Berlin, Brief 42, pp 351–365, 367–371
Michell J (1751) A treatise of artificial magnets, 2nd edn. Joseph Bentham, Cambridge
Mols R (1986) Gottignies, G. F. de. Dictionnaire d'histoire et de géographie ecclésiastiques 21:122–125 (coll. 921–924)
[Montucla JE] (1760) Philosophiae naturalis theoria, etc. [Review of]. J étranger, février, 52–74; mars, 61–88
Most GW (1984) Zur Entwicklung von Leibniz's Specimen Dynamicum. In: Heinekamp A (ed) Leibniz's dynamica. Studia Leibnitiana, Sonderheft 13, pp 148–163
Motta F (2001) I criptocopernicani. Una lettura del rapporto fra censura e coscienza intellettuale nell'Italia della Controriforma. In: Montesinos J, Solís C (eds) Largo campo di filosofare. Eurosymposium Galileo 2001. Fundación canaria orotava de historia de la ciencia, La Orotava, pp 693–718
Nanni S (2014) Passionei, Domenico Silvio. In: Dizionario Biografico degli Italiani (81: 000-000). Istituto della Enciclopedia italiana, Roma
Nedelkovitch [Nedeljković] D (1922) La philosophie naturelle et relativiste de R.-J. Boscovich. Éditions de la vie universitaire, Paris
Newton I (1687) The principia: mathematical principles of natural philosophy (trans: Cohen IB et al). University of California Press, Berkeley (1999)

Newton I (1706) Optice: Sive De Reflexionibus, Refractionibus, Inflexionibus & Coloribus Lucis. Libri Tres, Londinii: Impensis Sam. Smith & Benj. Walford, Regiæ Societatis Typograph. ad Insignia Principis in Cœmeterio D. Pauli
Newton I (1730) Opticks: or, a treatise of the reflections, refractions, inflexions and colours of light. Also two treatises of the species and magnitude of curvilinear figures. Printed for William Innys at the West-End of St. Paul's, London
Newton I (1962) In: Rupert Hall A, Boas Hall M (eds) Unpublished scientific papers: a selection from the Portsmouth Collection in the University Library, Cambridge. Cambridge University Press, New York
Newton I (2004) In: Janiak A (ed) Philosophical writings. Cambridge University Press, Cambridge
Olson R (1969) The reception of Boscovich's ideas in Scotland. Isis 60(1):91–103
Palmerino CR (2003) Two Jesuit responses to Galileo's science of motion: Honore Fabri and Pierre le Cazre. In: Feingold M (ed) The new science and jesuit science: seventeenth century perspectives. Springer, Dordrecht, pp 187–227. https://doi.org/10.1007/978-94-017-0361-1_4
Paoli G (1988) Ruggiero Giuseppe Boscovich nella scienza e nella storia del '700. Accademia Nazionale delle Scienze detta dei XL, Roma
Papineau D (1977) The Vis Viva controversy: do meanings matter? Stud Hist Philos Sci 8 (2):111–142
Peacock G (1834) Report on the recent progress and present state of certain branches of analysis. In: Report on the third meeting of the British Association for the advancement of science; held at Cambridge in 1833. John Murray, London, pp 185–352
Pepe L (2010) Introduzione. In: Pepe L (ed) ENo, II, Elementa Universae Matheseos. Elementi di geometria, pp 11–28
Peters CAF (1848) Recherches sur la parallaxe des étoiles fixes. Imprimerie de l'Académie Impériale des Sciences, St.-Pétersbourg
Proverbio E (2003) Gli interessi scientifici di Ruggiero G. Boscovich per i fenomeni elettrici e i suoi incontri con Benjamin Franklin ed altri elettricisti inglesi e francesi. Quaderni di Storia della Fisica 11:3–48
Quinn A (1982) Repulsive force in England, 1706-1744. Hist Stud Phys Sci 13(1):109–128
Reeves EA (1997) Painting the heavens: art and science in the age of Galileo. Princeton University Press, Princeton
Reiff RA (1889) Geschichte der unendlichen Reihen. Laupp, Tübingen
Riccioli GB (1651) Almagestum novum astronomiam veterem novamque complectens, 2 vols. ex Typographia haeredis Victorij Benatij, Bononiae
Rossi P (1999) Le sterminate antichità e nuovi saggi vichiani. La Nuova Italia, Firenze
Rowning J (1753) A compendious system of natural philosophy. Printed for Sam. Harding, London
Schaffer S (1980) Natural philosophy. In: Rousseau GS, Porter R (eds) The ferment of knowledge. Studies in the historiography of eighteenth-century science. Cambridge University Press, Cambridge, pp 55–91
Schneider I (1993) Johannes Faulhaber 1580-1635. Rechenmeister in einer Welt des Umbruchs. Basel, Birkhäuser
Schofield RE (1970) Mechanism and materialism. British natural philosophy in an age of reason. Princeton University Press, Princeton
Schofield RE (1997) The enlightenment of Joseph Priestley. Pennsylvania State University Press, University Park
Schubring G (2005) Conflicts between generalization, rigor, and intuition: number concepts underlying the development of analysis in seventeenth-nineteenth century France and Germany. Springer, New York
Schüling H (1969) Die Geschichte der Axiomatischen Methode im 16. und Beginnenden 17. Jahrhundert. Georg Olms, Hildesheim
Scott JF (1938) The mathematical work of John Wallis. Taylor & Francis, London (Reprint New York: Chelsea, 1981)

Sedley DN (2004) On generation and corruption I 2. In: de Haas FAJ, Mansfeld J (eds) Aristotle, on generation and corruption book 1. Symposium Aristotelicum. Oxford University Press, Oxford, pp 65–89
Sgarbi M (ed) (2010) Francisco Suárez and his legacy: the impact of Suárezian metaphysics and epistemology on modern philosophy. Vita e Pensiero, Milano
Shapiro AE (1993) Fits, passions, and paroxysms: physics, method, and chemistry and Newton's theories of colored bodies and fits of easy reflection. Cambridge University Press, Cambridge
Shields C (2012) Shadows of beings. In: Hill B, Lagerlund H (eds) The philosophy of Francisco Suárez. Oxford University Press, Oxford, pp 57–74
Siebert H (2005) The early search for Stellar Parallax: Galileo, Castelli, and Ramponi. J Hist Astron 36(3):251–271. https://doi.org/10.1177/002182860503600301
Siebert H (2006) Die große kosmologische Kontroverse: Rekonstruktionsversuche anhand des Itinerarium exstaticum von Athanasius Kircher SJ (1602–1680). Franz Steiner, Stuttgart
Siebert H (2007) La 'découverte' de la parallaxe stellaire par Robert Hooke—contexte et conséquences d'une experience. Cronos 10:41–61
Simmons A (1999) Jesuit Aristotelian education: the De anima commentaries. In: O'Malley JW, Bailey GA, Harris SJ, Kennedy TF (eds) The Jesuits: culture, learning and the arts, 1540-1773. Toronto University Press, Toronto, pp 522–537
Siret A (1884) Gottignies (Gilles-François de). In: Biographie nationale de Belgique (8: coll. 154–156). Bruylant-Christophe et Cie, Bruxelles
Sommervogel C (1890-1932) Bibliothèque de la Compagnie de Jésus, 11 voll., Bruxelles: Schepens. Picard, Paris
Spencer JB (1967) Boscovich's theory and its relation to Faraday's researches: an analytic approach. Arch Hist Exact Sci 4(3):184–202
Stan M (2017) Euler, Newton, and foundations for mechanics. In: Schliesser E, Smeenk C (eds) The Oxford handbook of Newton. Oxford University Press, Oxford. https://doi.org/10.1093/oxfordhb/9780199930418.013.31
Stay B (1744) Philosophiae versibus traditae libri VI. Apud Sebastianum Coleti, Rome (2nd revised edition Roma: ex Typographia Palladis)
Stay B (1755) Philosophiae recentioris versibus traditae libri X, cum adnotationibus, et supplementis P. Rogerii Josephi Boscovich, Tomus I. Palearini, Roma (Tomes II and III were published in 1760 and 1792, respectively)
Stedall JA (2004) [Introduction.] The arithmetic of infinitesimals. In: Wallis J (ed) The arithmetic of infinitesimals. Springer, New York, pp xi–xxxiv
Stein H (2002) Newton's metaphysics. In: Cohen IB, Smith GE (eds) The Cambridge companion to Newton. Cambridge University Press, Cambridge
Strazzoni A (2019) Burchard de Volder and the age of the scientific revolution. Springer, Cham
Struik DJ (1969) A source book in mathematics. Harvard University Press, Cambridge
Suardi G (1752) Nuovi istromenti per la descrizione di diverse curve antiche e moderne. Gian-Maria Rizzardi, Brescia
Tacconi I (1994a) I poemi filosofici latini di Benedetto Stay, il Lucrezio ragusino. In: Tacconi V (ed) Per la Dalmazia con amore e con angoscia. Tutti gli scritti editi ed inediti di Ildebrando Tacconi. Del Bianco, Udine, pp 333–374
Tacconi I (1994b) Benedetto Stay e il suo poema newtoniano. In: Tacconi V (ed) Per la Dalmazia con amore e con angoscia. Tutti gli scritti editi ed inediti di Ildebrando Tacconi. Del Bianco, Udine, pp 375–410
Taton R, Wilson C (1989) The general history of astronomy, vol 2. Planetary astronomy from the renaissance to the rise of astrophysics. Part A: Tycho Brahe to Newton. Cambridge University Press, Cambridge
Thackray A (1970) Atoms and powers: an essay on Newtonian matter-theory and the development of chemistry. Harvard University Press, Cambridge

Tho T (2017a) As matter to form so passive to active? The irreducible metaphysics of Leibniz's dynamics. In: Strickland L, Vynckier E, Weckend J (eds) Tercentenary essays on the philosophy and science of Leibniz. Palgrave Macmillan, Cham, pp 131–158

Tho T (2017b) Vis Vim Vi: declinations of force in Leibniz's dynamics. Springer, Cham

Tropfke J (1980) Geschichte der Elementarmathematik, Bd. 1. Arithmetik und Algebra. 4. Auflage. de Gruyter, Berlin

Truesdell C (1968) Essays in the history of mechanics. Springer, Berlin

Ugaglia M (2012) Due variazioni sul tema della potenza. In: Aristotele F (ed) Libro III, introduzione, traduzione e commento di Monica Ugaglia. Carocci, Roma

Ullmaier H (2005) Puncta, particulae et phaenomena: Roger Josef Boscovich und seine Naturphilosophie. Wehrhahn, Hannover

Valleriani M (2017) The tracts on the *sphere*: knowledge restructured over a network. In: Valleriani M (ed) The structures of practical knowledge. Springer, Cham, pp 421–473

van de Vyver O (1980) L'école de mathématiques des jésuites de la province flandro-belge au XVIIe siècle. Arch Hist Soc Iesu 49:265–278

van der Waerden BL (1983) Eulers Herleitung des Drehimpulssatzes. In: Fellmann EA (ed) Leonhard Euler 1707–1783. Beiträge zu Leben und Werk. Birkhäuser, Basel, pp 271–282

van der Waerden BL (1985) A history of algebra: from al-Khwārizmī to Emmy Noether. Springer, Berlin

van Looy H (1981) Sancto Vincentio, Gregorius a. In: Nationaal Biografisch Wordenboeck. Paleis der Academiën, Brussel, vol 9, pp 677–684

van Lunteren FH (1991) Framing hypotheses. Conceptions of gravity in the 18th and 19th centuries. PhD Thesis at Rijksuniversiteit, Utrecht

van Strien M (2017) Continuity in nature and in mathematics: Du Châtelet and Boscovich. In: Massimi M, Romeijn JW, Schurz G (eds) EPSA15 selected papers. European studies in philosophy of science, vol 5. Springer, Cham, pp 71–81

Vanpaemel GHW (2003) Jesuit science in the Spanish Netherlands. In: Feingold M (ed) Jesuit science and the republic of letters. MIT, Cambridge, pp 389–432

Varignon P (1706) Reflexions sur les Espaces plus qu'infinis de M. Wallis. Hist Acad R Sci Mém 1706:15–23

Villoslada RG (1954) Storia del Collegio Romano. apud aedes Universitatis Gregorianae, Roma

Virgil (1999) Eclogues. Georgics. Aeneid 1.-6, with an English translation by H. Rushton Fairclough, revised by G. Goold. Harvard University Press, Cambridge

Waff C (1975) Alexis Clairaut and his proposed modification of Newton's inverse-square law of gravitation. In: Avant, Avec, Après Copernic: La représentation de l'Univers et ses conséquences épistémologiques. Centre International de Synthèse, XXXIe Semaine De Synthèse. Albert Blanchard, Paris, pp 328–335

Waff C (1976) Universal gravitation and the motion of the Moon's Apogee: the establishment and reception of Newton's inverse-square law, 1687-1749. Unpublished Dissertation. The Johns Hopkins University

Waismann F (1970) Einführung in das mathematische Denken. DTV, München

Wallace WA (1972) Causality and scientific explanation, 2 vols. University of Michigan Press, Ann Arbor

Wallace WA (1984) Galileo and his sources. The heritage of the Collegio Romano in Galileo's science. Princeton University Press, Princeton

Wallis J (1656) Arithmetica Infinitorum: Sive Nova Methodus Inquirendi in Curvilineorum Quadraturam, aliaq[ue] difficiliora Matheseos Problemata. Typis Leon, Oxonii. Lichfield Academiae Typography (Eng trans: The Arithmetic of Infinitesimals. Translated from Latin to English with an Introduction by Jacqueline A. Stedall. Springer, New York, 2004)

Wallis J (1699) Operum Mathematicorum Volumen Tertium. Ex Theatro Sheldoniano, Oxoniae

Weijers O (2013) In search of the truth. A history of disputation techniques from antiquity to early modern times. Turnhout, Brepols

Wepster S (2010) Between theory and observations: Tobias Mayer's explorations of lunar motion, 1751–1755. Springer, New York

Westfall RS (1971) Force in Newton's physics: the science of dynamics in the seventeenth century. Neale Watson, New York

Whyte LL (1961) Boscovich's atomism. In: Whyte LL (ed) Roger Joseph Boscovich. Studies of his life and work on the 20th anniversary of his birth. Allen & Unwin, London, pp 102–126

Williams LP (1965) Michael Faraday: a biography. Chapman & Hall, London

Youschkevitch A (1976) The concept of function up to the middle of the nineteenth century. Arch Hist Exact Sci 16:37–85

Index

L. Guzzardi, *Ruggiero Boscovich's Theory of Natural Philosophy*, Science Networks. Historical Studies 60, https://doi.org/10.1007/978-3-030-52093-9

The manufacturer's authorised representative in the EU is Springer Nature Customer Service Centre GmbH, Europaplatz 3, 69115 Heidelberg, Germany. If you have any concerns regarding our products, please contact ProductSafety@springernature.com

Printed and bound by CPI Group (UK) Ltd, Croydon, CR0 4YY
30/09/2026
02225492-0014